J. B.

Rays of Light for Dark Hours

Consolation for the Afflicted

J. B.

Rays of Light for Dark Hours
Consolation for the Afflicted

ISBN/EAN: 9783337253523

Printed in Europe, USA, Canada, Australia, Japan

Cover: Foto ©berggeist007 / pixelio.de

More available books at **www.hansebooks.com**

RAYS OF LIGHT

FOR

DARK HOURS.

By J. B.

WITH AN INTRODUCTION

By R. R. BOOTH, D. D.

NEW YORK:

ANSON D. F. RANDOLPH,

No. 770 BROADWAY.

1864.

INTRODUCTION.

The need of consolation is deeply seated in the life of man. It is true that the consciousness of his strength sometimes makes him unmindful of the exposure of his earthly condition, and he goes on his way, amid the forces of the world, holding his head erect, and in his heart defying harm or hindrance. But only for a time. The storms which beat upon the human world never fail to rush at last upon such a defiant front, and drive the sturdiest heart to seek some place of shelter. Says an old proverb: "If you will dig but deep enough, under all earth you will find water, and under all life you will find grief."

It is sooner or later the common experience of all, that "in the world ye shall have tribulation." Amid the boasts of progress, strength and skill which men pour forth, there is continually heard an undertone of grief and pain which discloses the wide fulfillment of the primal curse.

> "The air is full of farewells to the dying,
>   And moanings for the dead;
> The heart of Rachel for her children crying,
>   Will not be comforted."

It is not wise for us to overlook the fact that the progress and development of life is to the large proportion of our race a continual Apocalypse of suffering. On every side we see it, and at some appointed time it is revealed to us in our own experience.

Ties which we would make perpetual are rapidly dissolved by unseen strokes. Treasures which we fondly hoped to retain securely are torn from us in the twinkling of an eye. Faces which have smiled on us and gladdened us with their light and beauty are blighted by the frost of death, and must be quickly buried from our sight. Thus all around us the stern Tragedy passes to its consummation, and we go on our way, knowing that somewhere in the waste the Shadow sits and waits our coming.

If we realize this aspect of our life, we shall acknowledge readily that there is no work on earth so blessed as that which seeks to impart consolation to the sorrowing or troubled heart. The loftiest genius or the most fervent piety is never so well employed as when put to service through speech or song for those who weep.

It was in this conviction that the apostle Paul wrote those words of deep significance: "Blessed be God, ever the Father of our Lord Jesus Christ, the Father of mercies, and the God of all comforts, who com-

forteth us in all our tribulation, that we may be able to comfort them which are in any trouble by the comfort wherewith we ourselves are comforted of God."

And it is good cause for gratitude that, apart from the sacred Scriptures and " the consolation which is in Christ," there are so many utterances of uninspired lips which aim to lighten the pressure of affliction, and to reveal the use and sacredness of sorrow.

For no one tongue can speak the words of soothing which are adapted to all modes of grief; no one experience can compass the mighty range through which the power of suffering is realized by human hearts. Our griefs demand the expression of many tongues of various experiences. For the great work of consolation, the gold, the frankincense and the myrrh are all required; so that from the blended treasures of many minds there may be gathered the soothing or inspiring influence which will be adapted to some particular affliction.

It was in a sincere apprehension of these truths that this book of selections has been prepared for public use.

It lays claim to no originality save in the arrangement which aims to harmonize selections of prose and poetry, and to suit the want of every form of sorrow. It is emphatically a book for " those who are in any

*trouble," and is commended to those into whose hands
it may fall, with this intent alone.*

*The writer of these lines of introduction has him-
self found comfort and relief in turning the pages of
the manuscript, and it is partly at his solicitation that
it has been published. He cordially unites with the
compiler in the desire that it may avail to soothe some
wounded spirits, and may remind those who mourn
that though "weeping may endure for a night, joy
cometh in the morning."*   *R. R. B.*

New-York, July 25th, 1862.

Rays of Light for Dark Hours.

Rejoice, O grieving heart!
  The hours fly past ;
With each some sorrow dies,
With each some shadow flies,
  Until at last
The red dawn in the east
Bids weary night depart,
  And pain is past.
Rejoice, then, grieving heart,
  The hours fly past.

MISS PROCTOR.

"Yet man is born unto trouble, as the sparks fly upward." Such is the divine decree, and who can claim exemption from its operation ?

All events which affect our moral or spiritual interests are governed by a divine law as certainly as the changes which take place in the material world. Sin and death, " with all our woe," stand in the relation of cause and consequence, as truly as do the laws of gravitation to the motions of the heavenly bodies.

It is, however, an affecting evidence of God's loving kindness and tender mercy, that the sufferings of this life, the penalties of sin, are made a necessary part of our earthly discipline, and are not inflicted on his children, under the Gospel dispensation, as an indication of wrath; for we are kindly assured, that "whom the Lord loveth he chasteneth," and that our chastisements are for our " profit," — and oh! what a profit! — " that we might be partakers of his holiness."

The very first lesson that these truths should teach us, in the times of our trials and afflictions, is that of absolute submission to the will of God.

If you can now, my afflicted friend, when all appears so dark and desolate, say with all your heart in the words our Saviour has taught us, Thy will be done ; I trust you will find some consolation and comfort, some healing balm for a wounded spirit, in the following pages.

Submit we must to all of God's dealings with us, willingly or unwillingly. Those only who trace their afflictions to a Father's hand, will find that it is good to be afflicted; by such, some of "the peaceable fruits of righteousness," which our "afflictions," however "grievous," should produce, may be gathered in the extracts from various authors which I have collected for my own use; for I also have been a "stricken deer;" and as I have received comfort from their perusal, I hope they may be of service to others who as sons and daughters of sorrow, are called to sit in the mourners' seat.

J. B.

Rays of Light for Dark Hours.

THE GRAVE.

THE grave! the grave! it burys every error, covers every defect, extinguishes every resentment. From its peaceful bosom spring none but fond regrets and tender recollections. Who can look down upon the grave even of an enemy, and not feel a compunctious throb that he ever should have warred with the poor handful of earth that lies mouldering before him! But the grave of those we loved — what a place for meditation! There it is that we call up in long review the whole history of virtue and gentleness, and the thousand endearments lavished upon us almost unheeded in the daily intercourse of intimacy. There it is that we dwell upon the tenderness, the solemn, awful tenderness of the parting scene; the bed of death, with all its stifled griefs, its noiseless attendance, its mute, watchful assiduities, the last testimonies of expiring love, the feeble, fluttering, thrilling (oh! how thrilling!) pressure of the hand, the last fond look of the glazing eye turning upon us even from the threshold of existence, the faint, faltering accents struggling in death to give one more assurance of affection! Ay, go to the grave of buried love and meditate; there settle the account with thy conscience for every past benefit unrequited, every past endearment unregarded, of that being who can never, never, never return to be soothed by thy contrition! If thou art a child, and hast ever added a sorrow to the soul, or a furrow to the brow of an affectionate parent; if thou art a husband, and hast ever caused the fond bosom that ventured its whole happiness in thy arms to doubt one moment of thy kindness or thy truth; if thou art a friend, and hast ever wronged in thought, word, or deed, the spirit that generously confided in thee; if thou art a lover, and hast ever given one unmerited pang to that true heart that now lies cold and still beneath thy feet, then be sure that every unkind look, every ungracious word, every ungentle action, will come thronging back upon thy memory, and knocking dolefully at thy soul; then be sure that thou wilt lie down sorrowing and repentant on the grave, and utter the unheard groan, and pour the unavailing tear, more deep, more bitter, because unheard, unavailing.— *Washington Irving.*

YET mourn not for the just,
The loved—the lost—no tears recover them!
No sorrowing memory brings them from the dust!

STORMY TRIALS.

On! happy for us if all the hurricanes that ruffle life's unquiet sea have the effect of making Jesus more precious. If God has to employ stormy trials, severe afflictions for this end, let us not quarrel with the wise ordination. Better the storm WITH Christ than the smooth water WITHOUT him.

> " Far more the treacherous calm I dread
> Than tempests bursting overhead."

It is the experience not of the luxurious barrack, but of the tented field, the trench, and night-watch, which makes the better and hardier soldier. It is not the exotic nursed in glass and artificial heat which is the type of strength, but the plant struggling for existence on bleak cliffs, or the pine battling with Alpine gusts, or shivering amid Alpine snows. If there be a sight in the spiritual world more glorious than another, it is when one sees (as may often be seen) a believer growing in strength and trust in God by reason of his very trials; battered down by storm and hail, a great fight of afflictions—enduring loss of substance, loss of friends, loss of health, yet standing by emptied coffers and full graves, and with an aching but resigned heart, enabled to say : " Heart and flesh do faint and fail, but God is the strength of my heart and my portion forever."— *Macduff*.

> EVERY creature hope and trust,
>   Every earthly prop or stay,
> May be prostrate in the dust,
>   May have failed or passed away ;
> Yet a season tarry on—
>   Nobly borne is nobly done.

SICKNESS SANCTIFIED.

BETTER, however, than the most sanguine expectation of a cure, is the sanctified use of sickness. God has
different ways of making his children holy, but with
many it is his plan to make them perfect through sufferings. Baxter says in his note on the cure at Bethesda:
" How great a mercy was it to live thirty-eight years
under God's wholesome discipline! O my God! I thank
thee for the like discipline of fifty-eight years; how safe
is this in comparison of full prosperity and pleasure!"

We often recall what was once told us by a sainted
friend whose parish was the Grassmarket of Edinburgh,
that when wearied and sickened with the scenes of de-
pravity which he constantly encountered, before returning home for the day, he often went to refresh his spirit
in a garret where a poor woman was slowly dying of a
cancer. But so much of heaven had come down to that
little chamber, that just as in the peace of God the sufferer triumphed over nature's agony, so in sharing her
wonderful happiness, the man of God forgot the wicked
ness with which his soul had been vexed all day, as he
also forgot the deplorable misery of the tenement in
which this beatified spirit still lingered. Glad and
glorious infirmity, which secures the Saviour's presence,
and is sustained in the Saviour's power!—*Hamilton.*

> I WISHED a flowery path to tread,
>   And thought 'twould safely lead to heaven;
> A lonely room, a suffering bed,
>   These for my training-place were given.

Long I resisted, mourned, complained,
Wished any other lot my own ;
Thy purpose, Lord, unchanged remained—
What wisdom planned love carried on.

VOICES FROM THE GRAVE.

It is indeed the rule of life generally that no man profits by any experience but his own ; yet there is one kind of experience which may perhaps be considered an exception. It is that gained at the death-beds of those who " die in the Lord." Few persons probably have attained mature age without having had some experience granted to them.

They who have gone before us in suffering, they whose footsteps we have followed with sympathy, have left with us a blessing which they little thought of—a strength even in the very spectacle of their weakness. All those hours of lingering pain at which we so wondered, asking, perhaps, in moments of unbelief, whether God could indeed love those whom he so afflicted, were hours of untold value, for they were tracing the record of that mighty strength by which the saints of God are enabled to wait with patience the appointed time " till their change come." All the words and looks of faith and love were prophecies and promises of the spirit of faith and love, who will be at hand when we need his aid. The gradual lessening of these earthly cares which made us marvel as we watched the change that passed over them, the calm acquiescence in God's will, the bright hope, the present realization of future happiness

—they were all treasures gathered for our use, which no effort could have purchased, no gladness of this world could have procured us. Therefore are these memories infinitely precious.—*Sewell.*

WE, too, have
    Been as thou art.

Tossed on the troubled waves,
    Life's stormy sea;
Grief and change manifold,
    Proving like thee.

Hope-lifted, doubt-depressed,
    Seeing in part ;
Tried, troubled, tempted,
    SUSTAINED as thou art.

Our God is thy God—what he
    Willeth is best ;
Trust him as we trusted, then
    Rest as we rest.

HE IS DEAD!

IT is long before we become assured, as it were, of the loss of those we value. Vague and imperfect as our ideas of that terrible separation, are the first feelings which attend it. We grieve, indeed; but while we grieve there is a want of reality and certainty in our sorrow. We repeat to ourselves that they are lost, gone, vanished forever, and even while we repeat it, feel as though they might return. For months the possibility of writing to them lingers vaguely in our minds ; they seem absent, not *buried*. We recollect that they are dead with a burst of weeping. It is not till long seasons have revolved, till joys which they would have shared, anxieties which they might have alleviated, events in which they would have their part, have all been our portion and ours only; till the grasp of welcome or congratulation has been long unfelt, till the opinions we used to value have been long unasked, till we have stood in some trial of life, and felt the want of our accustomed counsellor and friend, that we thoroughly comprehend the world of separation, and bereavement contained in that short phrase: "He is dead!"—*Mrs. Norton.*

> TIME hath not power to bear away
>   Thine image from my heart;
> No scenes that mark life's onward way
>   Can bid it hence depart.
>
> Amid earth's conflict, woe, and care,
>   Where our dark path appears,
> 'Tis sweet to know thou canst not share
>   Our anguish or our tears.
>
> Yet while our souls, with anguish riven,
>   Mourn, loved and lost, for thee,
> We raise our tearful eyes to heaven,
>   And joy that thou art free.

BUNYAN'S TRIALS.

I FOUND myself a man encompassed with infirmities. The parting with my wife and poor children hath often been to me in this place as the pulling the flesh from the bones, and that not only because I am somewhat too fond of these great mercies, but also because I should have often brought to my mind the many hardships, miseries, and wants that my poor family were likely to meet with, should I be taken from them, especially my poor blind child, who lay nearer my heart than all beside. Oh! the thoughts of the hardships I thought my poor blind one might go under, would break my heart to pieces. Poor child! thought I, what sorrow art thou like to have for thy portion in this world! Thou must be beaten, must beg, suffer hunger, cold, nakedness, and a thousand calamities, though I can not now endure the wind should blow upon thee. But yet, recalling myself, thought I, I must venture you all with God, though it goeth to the quick to leave you.— *Bunyan.*

> Shall I not trust my God,
> Who doth so well love me?
> Who as a father cares so tenderly?
> Shall I not lay the load
> Which would my weakness break
> On his strong hand, who never doth forsake?
>
> Who doth the birds supply?
> Who grass and trees and flowers?
> Doth beautifully clothe through ceaseless hours?
> Who hears us ere we cry?
> Can he my need forget?
> Nay, though he slay me, I will trust him yet.

AGNES.

"WITH patience then the course of duty run ;
God never does or suffers to be done,
But that which you would do if you could see
The end of all events as well as he."

"THE thought in those lines has done more to sustain me, or at least to keep my mind quiet, than any uninspired words."

"No doubt it is literally true," said I, "that if we could have seen all which God saw, we should have said: 'How desirable it is that Agnes should die now.' We never would have taken the responsibility of judging, however ; and therefore it is well that there is One who can and who is willing to do so, and does not spare for our crying."

"What are some of the reasons," said she, "which you can imagine why it was best ?"

"Oh ! she might have had the seeds of disease in her, which would have made her life a burden," I replied.

"Or she might have proved a great trial to us in some way," she added.

"Perhaps," said I, "God wishes to prepare us to do great good in the world, and this is the preparative. If God seeks to fill us with himself, if he desires our love, what an honor it is and what a privilege it is to receive him, even by displacing the dearest object."—*Nehemiah Adams.*

IN that great cloister's stillness and seclusion
By guardian angels led,
Safe from temptation, safe from sin's pollution,
She lives, whom we call dead.

Let us be patient—these severe afflictions
　Not from the ground arise;
But oftentimes celestial benedictions
　Assume this dark disguise.

We see but dimly through the mists and vapors,
　Amid these earthly damps:
What seem to us but sad funereal tapers,
　May be heaven's distant lamps.

MY MOTHER'S GRAVE.

It was thirteen years since my mother's death, when after a long absence from my native village, I stood beside the sacred mound beneath which I had seen her buried. Since that mournful period, a great change had come over me. My childish years had passed away, and with them my youthful character. The world was altered too, and as I stood at my mother's grave I could hardly realize that I was the same thoughtless, happy creature whose cheeks she so often kissed in an excess of tenderness. But the varied events of thirteen years had not effaced the remembrance of that mother's smile. It seemed as if I had seen her but yesterday—as if the blessed sound of her well-remembered voice was in my ear. The gay dreams of my infancy and childhood were brought back so distinctly to my mind, that had it not been for one bitter recollection, the tears I shed would have been gentle and refreshing. The circumstance may seem a trifling one, but the thought of it now pains my heart, and I relate it that those children who have parents to love them may learn to value them as they ought.

My mother had been ill a long time, and I had become so accustomed to her pale face and weak voice, that I was not frightened at them as children usually are. At first, it is true, I sobbed violently, but when, day after day, I returned from school and found her the same, I began to believe she would always be spared to me. But they told me she would die. One day when I had lost my place in the class, and done my work wrong side outward, I came home discouraged and fretful; I went to my mother's chamber. She was paler than usual, but she met me with the same affectionate smile that always welcomed my return. Alas! when I look back through the lapse of thirteen years, I think my heart must have been stone not to have been melted by it. She requested me to go down-stairs and bring her a glass of water. I pettishly asked why she did not call a domestic to do it. With a look of mild reproach, which I shall never forget if I live to be a hundred years old, she said: "And will not my daughter bring a glass of water for her poor sick mother?"

I went and brought her the water, but I did not do it kindly. Instead of smiling and kissing her as I was wont to do, I set the glass down very quickly and left the room. After playing a short time I went to bed without bidding my mother good-night; but when alone in my room, in darkness and silence, I remembered how pale she looked, and how her voice trembled as she said: "Will not my daughter bring a glass of water to her poor sick mother?" I couldn't sleep. I stole into her chamber to ask forgiveness. She had sunk into an easy slumber and they told me I must not waken her. I did not tell any one what troubled me, but stole back to my bed, resolved to rise early in the morning, and tell her how sorry I was for my conduct.

The sun was shining brightly when I awoke, and hurrying on my clothes, I hastened to my mother's chamber. She was dead! She never spoke more—never smiled upon me again; and when I touched the hand that used to rest upon my head in blessing, it was so cold that it made me start. I bowed down by her side and sobbed in the bitterness of my heart. I thought then I wished I might die and be buried with her; and old as I now am, I would give worlds were they mine to give, could my mother but have lived to tell me she forgave my childish ingratitude. But I can not call her back; and when I stand by her grave, and whenever I think of her many kindnesses and love, the memory of that reproachful look she gave me still "bites like a serpent and stings like an adder."—*Anonymous.*

MY spirit yearns to bring
The lost ones back—yearns with desire intense,
And struggles hard to wring
Thy bolts apart, and pluck thy captives thence !

They have not perished—no !
Kind words, remembered voices, once so sweet—
Smiles radiant long ago,

.   .   .   .   .   .

SAFE IN THE FOLD.

AND yet how much better for my lamb to be suddenly housed, to slip unexpectedly into the fold to which I was conducting her, than remain exposed here! Perhaps to become a victim! I cried: " O Lord! spare my child !" He did, but not as I meant ; he snatched it from danger, and took it to his own home.

When I pass by the blaze of dissipation and intemperance, I feel a moment's relief. I say to my heart, " Be still ;" at least she is not left to follow these IGNES FATUI. How much better is even the grave for my child than the end of these things? Help me, O my God and Father! to recollect that I received this drop of earthly comfort from a spring which still remains! Help me to feel that nothing *essential* is altered, " for with thee is the fountain of life !" Part of myself is already gone to thee: help what remains to follow.— *Richard Cecil.*

> THY gourd has fallen ! Yet had its pleasant shade
> Been spared for future years to bless thy bower,
> It would have lived, but only to decay !
> Those bursting buds and blossoms, early plucked,
> (Say not TOO early,) would at last have dropped
> As withered flowers. Let the great Husbandman
> Select the time to take his own, before
> The chilling frosts of life have nipped it.
>   'Tis the exotic
> Which has been taken to a kindlier soil,
> To bloom unfading in far happier climes,
> Where tempests are unknown. Think of the storms
> That tender sapling has in love been spared.

THE LIVING LOST.

FAR happier the mother of the dead than the mother
of the reprobate. Happy those in whose cup, if there
is bitter sorrow, there is not also burning shame, and
who, in the day of their sore calamity, are spared the
agony of crime! You may have a child or dear relation
who is like to bring your gray hairs with sorrow to the
grave. And what are you to do? It seems as if nothing
could stop him in his wild career. He seems as if he
could not stop himself. He really looks as if he were
possessed with the devil. You have got good people
to talk to him, and you have talked to him yourself.
But it was of no use. He did not stop his ears; but
as for giving you any hold on his heart, his will,
you might have been a thousand miles away. And now
you have entirely lost sight of him. You know not
where he is, and what are you to do? Why this: you
have heard of the "fame" of Jesus: go to him and take
your child, your husband, your lost friend with you.
Take him, that is, as the nobleman and the woman took
their child. Take him in the arms of believing and importunate intercession. Say: "Thou Son of David, have
mercy upon me." He is the enemy of God and his
own soul. He is the slave of divers lusts and passions.
Thou knowest our frame. Thou knowest the affection
I feel for him. Thou knowest the faith I have in thee.
Oh! that Ishmael might live before thee! Oh! that
this wanderer may be restored, this madman brought
to his right mind! I know not where he is. At this

very moment thou compassest his path, and art acquainted with all his ways. Thou who hast the keys of David canst open for thyself that door : even now his heart is in thy hand. Oh ! speak the word and add a heaven to my heaven, a jewel to thy crown.—*Hamilton.*

> Yet there are pangs of deeper woe,
>   Of which the sufferers never speak,
> Nor to the world's cold pity show
>   The tears that scald the cheek,
> Wrung from their eyelids by the shame
> And guilt of those they shrink to name,
> Whom once they loved with cheerful will,
> And love, though fallen and branded, still.

DEEP WATERS.

How often does God hedge up our way with thorns to elicit simple trust! How seldom can we SEE all things so working for our good! But it is better discipline to BELIEVE it. "A great deep" is all the explanation thou canst often give to his judgments ; the *why* and the *wherefore* he seems to keep from us, to test our faith, to discipline us in trustful submission, and lead us to say : "Thy will be done." What are called "dark dealings" are the ordinations of undeviating faithfulness. Man may err, his ways are often crooked, "but as for God, his way is perfect." "He keepeth the feet of his saints." He leads sometimes darkly, sometimes sorrowfully, but most frequently by cross and circuitous ways we ourselves would not have chosen ; but always wisely, always tenderly. With all its mazy windings and turnings, its roughness and ruggedness, the believer's is not only A right way but THE right way, the best which covenant love and wisdom could select. Every individual believer, the weakest, the weariest, the faintest, claims his attention. His loving eye follows me day by day out to the wilderness, marks out my pasture, studies my wants, and trials, and sorrows, and perplexities, every steep ascent, every brook, every winding path, every thorny thicket. It is not rough driving, but gentle guiding.—*Macduff*.

THE way seems dark about me ; overhead
The clouds have long since met in gloomy spread ;
And when I looked to see the day break through,
Cloud after cloud came up with volume new.

And in that shadow I have passed along,
Feeling myself grow weak as it grew strong,
Walking in doubt and searching for the way,
And often at a stand, as now, to-day.

Perplexities do throng upon my sight,
Like scudding fog-banks to obscure the light ;
Some new dilemma rises every day,
And I can only shut my eyes and pray !

"MINE OTTO."

In a miserable old frame-house, so open that the snow and rain without difficulty found its way in, a Prussian mother and her children were striving to make themselves comfortable. Her children numbered three, all

of them boys — about eleven, four, and two years of
age. Their father had been dead but a few months.
A few shillings and the smallest quantity of furniture
were all the poor man left. When the father was dead,
the mother purchased a small stock of thread, needles,
pins, and tapes, and with the youngest child went from
door to door, from morning till night, leading the little
fellow till his legs would give out, and then he must be
carried; and thus hours every day would this devoted
mother bear about on one arm her basket, and on the
other this heavy child. The poor widow told me how
they came to America, and how happy they were till her
husband died; and when he died, how dark every thing
seemed; and it was night yet, with scarcely a gleam of
light. They suffer for want of food, raiment, and a
comfortable tenement. We spoke of her parting with
the children, and this seemed to add so much to her
already deep sorrow, that we could not urge it.

I said to the oldest boy: "How would you like to
have me get you a place in the country?"

Hesitating a little while, he turned with a smile to
me, his eyes swimming with tears, and answered:
"Me can no leave me modder."

This boy was in the habit of leading in prayer, morning and evening, with his mother and his little broth
ers, and there seemed so much affection on his part
toward them, and such a disposition to do what he
could to help his mother, it appeared cruel to separate
them. A representation of their case to the Society
procured them a supply of garments and bedding
enough to make them comfortable. The mother's

thanks for these favors, in broken English, were very
emphatic. In a short time we succeeded in getting
Otto a situation as a messenger or errand-boy, for
which he received twelve shillings a week. It was
found, however, after a few weeks, that his limited
knowledge of the language unfitted him for the place;
and his employer, paying him twelve shillings more
than was due him, sent him to us with a note stating
the fact. In less than a week we obtained a situation
in a tin-man's shop, where he received TEN dollars a
month. And now they felt rich, indeed. While yet
enjoying this new turn in their fortune, they were
awakened at midnight to escape only with their lives
from their burning dwelling. An adjoining carpen-
ter's shop had been set on fire, and communicated the
flames to their abode, destroying with it their little all.
Another tenement was hired; Otto was clothed from
the Home; so that he was away from his work but one
day on account of the fire. . . . For more than a
year every thing seemed prosperous; then a darker
cloud than almost any previous one came with blinding power and quickness over this poor stranger's soul.
She came to us one Saturday morning, the very picture
of woe, wringing her hands, and exclaiming almost as
soon as she saw us: "O mine Got! Mr. Halliday,
mine poor Otto! mine poor Otto!" and then with a
kind of wailing cry, she sat for some moments, and
seemed utterly heart-broken, I hardly daring to ask
her a question, so utterly crushed did she seem.

In answer to inquiry for particulars, she said: "He
say so pleasant, when he went to his work, 'Good

morning, mother;' but he no come back to say, 'Good evening!'" And then she again sat and cried aloud, until we asked once more to be made acquainted with the facts.

She said he went away early on Friday morning to his work, but did not come home, as usual, at evening. She waited for him until it was quite late, and then she was so troubled, getting some of the neighbors to take care of the little children, she started from their house, on Thirty-seventh street, to go to his shop, in the vicinity of the Astor House — a distance of more than three miles. Finding the shop shut, she turned toward Broadway, and on the corner of Ann street inquired of a policeman if he had seen such and such a boy. He told her that a boy was run over that morning in front of the Astor House, and that he was at the station-house on Warren street, and they started to go there; "and," to use her own language, "down in the cellar I found mine poor dead Otto!"

She had come to me in her trouble. The boy still lay at the station-house. We had it removed to the house which he had left so pleasantly only the day before, and we made arrangements for his interment in a rural cemetery, a few miles from the city. On Sunday morning, at an early hour, with a few of their countrymen and the kind Englishwoman who had at first directed our attention to them, we said a few words to the simple gathering, and lifted up our prayer for the widow and orphans to the widow's God, and then her "poor dead Otto" was carried out to sleep in the country burying-ground.

More than eight months have passed since we lifted
this almost frantic mother from her knees — her arms
clinging to the coffin of her dead Otto—yet her sorrow
seems as fresh as if it were but yesterday. She will sit
and speak so touchingly and tenderly of her buried
boy, and then her spirit is so chastened and so sweet,
I wished, as she this moment left my house, I could
have daguerreotyped her face, tones, and words. —
Halliday's Lost and Found.

THE spoiler hath come
    With his cold, withering breath,
And the loved and the cherished
    Lies silent in death !

And, oh ! do we question
    With tremulous breath,
Why the joy of your household
    Has fallen in death ?

Do you mourn round the place
    Of his perishing dust ?
Look onward and upward
    With holier trust.

THOUGHTS CONCERNING A DEPARTED FRIEND.

WHITHER is she gone? In what manner does she
consciously realize to herself the astonishing change?
How does she look at herself — as no longer inhabiting
a mortal tabernacle? In what manner does she recollect her state — as only a few weeks since? In what
manner does she think, and feel, and act, and communi-

cate with other spiritual beings? What manner of
vision has she of God and the Saviour of the world?
How does she review and estimate the course of discipline through which she had been prepared for the
happy place where she now finds herself? In what
manner does she look back on DEATH, which she has so
recently passed through? And does she plainly UN-
DERSTAND the nature of a phenomenon so awfully mysterious to the view of mortals? How does she remember and feel respecting us, respecting ME? Does she
indulge with delight a confident anticipation that we
shall, after a while, be added to her society? Earnest
imaginings and questionings like these arise without
end, and still, still there is no answer, no revelation.
The mind comes, again and again, up close to the thick
black vail; but there is no perforation, no glimpse.
She that loved me, and, I trust, loves me still, will not,
can not, must not, answer me. I can only imagine her
to say: "Come and see; serve our God, so that you
shall come and share at no distant time." One of the
most striking circumstances to my thought and feeling
is, that in devotional exercises, though she comes on
my mind in a more affecting manner than, perhaps,
ever, *I have no longer to pray for her.* By a momentary lapse of thought, I have been, I think, several
times on the point of falling into an expression for her,
as if still on earth; and the instant, "No; no more for
HER," has been an emotion of pain, and as it were, disappointment, till the thought has come: "*She* needs
not; she is now safe, beyond the sphere of mortals,
and their dangers and wants, in the POSSESSION of all

that prayer implored." Even AFTER this consolatory
thought, there has been a pensive trace of feeling,
something like pain, that sympathy, care for her wel
fare, should now be superfluous to her, and finally ex
tinguished.—*John Foster.*

I THINK of thee, when wintry storms are throwing
    Their snow-wrought shrouds around your dear old home;
Yet angel-voices give me gentle warning,
    To raise my thoughts to heaven, where thou art gone.
Thy vacant chair stands by our fireside still;
    Thy well-worn Bible rests upon my knee;
Importunate prayers rise to our Father still;
    But, oh! they are not for THEE—they are not for thee!

GO AND TELL JESUS.

Go and tell Jesus every thing. Tell him of your
bodily infirmities. Tell him of your waning health; of
your failing vigor; of your progressive disease; of the
pain, the lassitude, the nervousness, the weary couch,
the sleepless pillow, which no one knows but him.
Tell him of your dread of death—how you recoil from
dying — and how dark and rayless appears the body's
last resting-place.-- *Winslow.*

My Saviour! take from me now all vain regret;
Let me not mourn o'er hopes forever set;
O'er broken energies and prostrate life:
Am I not saved the toil, the jar, the strife?
And from my couch of pain to yonder sky,
How little intercepts the longing eye!
Docile of heart, and lowly may I be,
My Saviour! till I reach my home and thee.

TWO YEARS IN HEAVEN.

DEEM not these blossoms prematurely plucked.
No flower can drop too soon, if ripe for glory.
Early plucked is early bliss.
An early death-bed is an early crown.

Two years ago to-day he went to heaven. With us they have been long, long years since we heard the sound of his sweet voice, and the merry laugh that burst from his glad heart. He was the youngest of our flock. Three summers he had been with us, and, oh! he was brighter and sunnier than any summer day of them all. But he died as the third year of his life was closing. The others were older than he; and all we had of childhood's glee and gladness was buried when we laid him in the grave. Since then our hearts have been yearning for the boy that is gone. "Gone, but not lost," we have said a thousand times; and we think of him ever as living and blessed in another place not far from us.

Two years with CHRIST! It is joy to know that our child has been two years with the Saviour, in his immediate presence, learning of him, and making heaven vocal with songs of rapture and love. The blessed Saviour took little children in his arms when he was here on earth, and he takes them in his bosom there. Blessed Jesus! blessed children! blessed child!

Two years in heaven! They do not measure time in that world: there are no weeks, or months, or years; but all the time we have been mourning his absence here, he has been happy there. And when we think of what he has been enjoying, and the rapid progress he has been making, we feel that it is well for him that he has been taken away.

Two years with angels! They have been his constant companions, his teachers too; and from them he has drawn lessons of knowledge and love.

Two years with the redeemed! There are some among those redeemed who would have loved him here, had they been living with us; but they went to glory before him, and have welcomed him now to their company. I am not sure they know him as our child; and yet do we love to think that he is in the arms of those who have gone from our arms. And thus broken families are reünited around the throne of God and the Lamb.

He often wept when he was with us; he suffered much before he died; but now for two years he has not wept! And when we think of joys that are his, we are more than willing that he should stay where he now dwells, though our home is darkened by the shadow of his grave, and our hearts are aching all the time for his return. Long and weary have been the years without him; but they have been blessed years to him in heaven.—*S. I. Prime.*

> ANOTHER little form asleep,
>   And a little spirit gone;
> Another little voice is hushed,
>   And a little angel born.
>
> Two little feet are on the way
>   To the home beyond the skies;
> And our hearts are like the void that comes
>   When a strain of music dies.

A pair of little baby shoes,
And a lock of golden hair ;
The toy our little darling loved,
And the dress she used to wear.

The little grave in the shady nook,
Where the flowers love to grow;
And these are all of the little hope
That came three years ago.

DEATH OF A MOTHER.

"You have lost your child," said Mrs. Wales, "and you are not to leave her behind you. Some might think that you have more to be thankful for than I ; it may not seem so hereafter. When my six children come to me in heaven, having been useful here, bringing their sheaves with them, how glad I shall be that I had six orphans to trust to God !" "But yet," said my wife, "what sight is more heart-rending than a family of orphans ?" "Yes," said I, "but observation has led me to feel less and less solicitude on seeing a family of children left in orphanage by parents who were truly the children of God. The self-reliance, the restraining and subduing power of a deceased parent's memory, the friends raised up for them, all afford a good comment on these words : 'Leave thy fatherless children; I will preserve them alive.' Nothing seems to us more in violation of the natural and proper order of things, than the removal of a mother from a family of young children. We would

have provided against such a calamity by a special law,
had we arranged the affairs of life and death. He who
is willing to do so great and solemn a thing as to remove
a mother from the head of her family, must have reasons for it, as Mrs. Wales says, which would satisfy us
could we see them with a right mind. Such an event
is so peculiarly an act of God's providence, we may
suppose that He who giveth to the beast his food and
to the young ravens which cry, will not fail to accom-
plish some great and good purpose by it to all who love
him."—*Nehemiah Adams.*

YET would we say, what every heart approveth—
    Our Father's will,
Calling to him the dear ones whom he loveth,
    Is mercy still !

Not upon us or ours, the solemn angel
    Hath evil wrought ;
The funeral anthem is a glad evangel ;
    The good die not !

God calls our loved ones, but we lose not wholly
    What he has given ;
They live on earth, in thought and deed, as truly
    As in his heaven.

NO SICKNESS.

"THE inhabitant shall no more say, I am sick." Ye
who are now laid on beds of languishing and pain, listen to this. Now, as the shadows of each returning
evening begin to fall, you may have nothing but gloomy

anticipations. The morrow's light, which brings health and joy to a busy world, may bring nothing to you but fresh prostration and anguish. Meanwhile, as you lie tossing on your sick-bed, seek not to ask, "Am I getting the better of my pain?" but: "Am I made the better FOR it? Is it executing the great mission for which it has been sent of God? Is it sanctifying me, purging away the dross, and fitting me for glory?"— *Grapes of Eschol.*

> For all thy love bestows, I bless my lot;
> For all that love withholds, I murmur not;
> Sweet thoughts thou sendest to my solitude,
> And that which evil SEEMS from THEE is good;
> I ask thee not this sickness to remove;
> Only sustain me with thy pitying love!
> I ask not rest from weariness or pain,
> Only, Great Chastener, send them not in vain.
>
> Oh! wherefore heed this passing brief distress;
> A little suffering more, a little less,
> A little faltering through this checkered scene,
> And all will be as it had never been,
> Save that the burden of the weary road
> Led me to seek my strength in thee, my God!
> Save that the wish for ease, the hope of rest,
> Led me, my Father, to thy changeless breast.

THE HEREAFTER.

WHEN your father and myself enter on that great hereafter, then that will be a reality to you, which now seems so shadowy and uncertain. You love us, and I

know how often you will follow us in thought to the
mysterious abode, " in our Father's house " You will
wonder how we are occupied ; what our thoughts are
engaged about ; whether we love you still ; if we are
thoughtful about your present, and still anxious for
your future. And that strange, mysterious hereafter
will have a home aspect for you — you will expect to
receive a parent's welcome and have again a parent's
love. I am sure there will be in heaven the same strong,
tender love we always had for you here, but there will
be none of its corroding anxieties. I hope you will
continue to treasure up the pleasant memories of the
old homes we have had together here. And oh! I
know how often, when disappointments come, you will
long for " the wings of a dove," to fly to me for the sympathy and the love that never has failed you.—*A. N.*

> Through the mists of the hereafter,
>   In the land eternal dwelling ;
> Beyond the flood, the bitter flood of death,
>   Beyond the dark and turbid swelling
>   Of all earthly strife :
> They are waiting for us—watching,
> Watching, longing, hoping, waiting
>   In the Land Eternal.
>
> All who loved us—all our darlings,
>   Gone before us o'er the deep ;
> Moving through our lives as shadows,
>   Dim as visions in our sleep,
>   Live now the better life.
> We shall see their holy faces,
> We shall hear their loving voices
>   In the Land Eternal.

BEREAVEMENTS.

When death breaks in amongst our children, there is made a great gulf, and we, poor parents! can only look and feel and weep. The place well known amongst the rest is empty; the place at the table is empty; their place in your prayers is empty; and the face which met you at the door, with all its little news, meets you no more. Your little child was lovely, and singularly beloved. Be thankful that you had such a child. Be thankful that you had him so long. Be thankful that the Lord did not consult you how long the loan should be continued. His precious gifts might receive damage in our fond and foolish hands; for this cause the Father of mercies, in great tenderness, takes them and hides them from us, but at the same time lays them up, to be brought forth, and restored as a new source of great joy, at the meeting of the just men made perfect. — *John Jamieson.*

Bereaved mother! mourning o'er the loss
Of a departed child—a flower soon plucked,
(But not too soon for glory,) which distilled
Celestial fragrance on thy path below—
Weep not! but let thy envied boast be this:
"I am the parent of a ransomed saint."

WALKING IN DARKNESS.

It reminds us of the period of *soul*-darkness which sometimes overtakes the Christian pilgrim. "My serv-

ant that walketh in darkness and hath no light," says
God. Observe, he is still God's servant. He is the
"child of the light," though walking in darkness.
Gloom spreads its mantle around him—a darkness that
may·be felt. Shadows thicken upon his path. God's
way with him is in the great deep. "Thou art a God
that HIDEST thyself," is his mournful prayer. The Holy
Spirit is, perhaps, grieved; no visits from Jesus make
glad his heart; he is brought in some small degree into
the blessed Saviour's experience : "My God, my God,
why hast thou forsaken me?" But, sorrowful pilgrim,
there is a bright light in this your cloud — turn your
eyes toward it — the darkness through which you are
walking is not JUDICIAL. It is not the darkness of an
unconverted, alienated state. Oh! no; you are still a
"child of the day," though it may be temporary night
with your spirit. You are still a child, and God is still
a Father. "In a little wrath I hid my face from thee
for a moment ; but with everlasting kindness will I
have mercy on thee, saith the Lord, thy Redeemer."
"Is Ephraim my dear son ? is he a pleasant child ? for
since I spake against him I do earnestly remember him
still."—*Octavius Winslow.*

> GOD doth not leave his own :
> The night of weeping for a time may last ;
> Then, tears all past,
> His going forth shall as the morning shine,
> The sunshine of his favor shall be thine ;
> God doth not leave his own.
>
> God doth not leave his own :
> Though few and evil all their days appear ;

Though grief and fear
Come in the train of earth and hell's dark crowd,
The trusting heart says, even in the cloud,
God doth not leave his own.

THE FURNACE.

I OFTEN feel like a sacrifice. However, Jesus will take care that his Father is glorified, in spite of all our crying while the rod is in his hand. That thought often comforts me. And I was thinking this week that it is really a privilege to be in his furnace at all; for it is not intended for reprobate silver, but only for choice gold; and if we were not his choice gold, we should not have been put in there.—*Adelaide Newton.*

FEAR thou not then this furnace, for He lights it,
Not to destroy, but only to refine;
To purify the gold, and purge away
The dross, and fit for glory. Wondrous thought,
The great Refiner, seated by the fires,
Tempering their fury.

AN INFANT IN HEAVEN.

"SHE is ours still. She may have ten thousand instructors in heaven, but we are her parents. It seems to me a great honor to be a parent of a redeemed soul. How much nearer this brings us to a likeness with God than angels approach! She is our precious child still. Her past history, the memory of her, the happiness she afforded us, the love to each other of which she was the occasion, the beautiful, hallowed thoughts which we shall continue to have about her, are a possession which no one can take from us. She was God's gift, and she is ours still. You asked me, when we came from the funeral, whether I regretted all the sickness and sorrow which Agnes cost. To have a child in heaven is worth all that a parent can suffer."—*Nehemiah Adams.*

Thou bright and star-like spirit!
    That in my visions wild,
I see 'mid heaven's seraphic host—
    Oh! canst thou be my child?

My grief is quenched in wonder,
    And joy arrests my sighs—
A branch from this unworthy stock
    Now blossoms in the skies.

The little weeper—tearless!
    The sinner—snatched from sin!
The babe to more than manhood grown
    Ere childhood did begin.

What bliss is born of sorrow!
    'Tis never sent in vain;
The heavenly Surgeon maims to save;
    He gives no useless pain.

DEATH OF THE FIRST-BORN.

During the days of his illness in Beckenham, Thomas Ward had been looking forward with deep delight to

the prospect of being admitted three times in the course
of the month of May to partake of the Sacrament of
the Lord's Supper ; understanding that it would be ad-
ministered on Ascension-Day, Whit-Sunday, and Trinity
Sunday. It had been a source of sacred joy to us both
to speak together of these opportunities of confessing
his faith in Christ publicly.

We little thought that before the earliest of the appointed days came, he would be leaning — like the beloved disciple, as he drank of the cup of the Last Supper — on the bosom of his Saviour ; in tranquil and
blessed anticipation of the hour when the Lord Jesus
shall " drink it new" with all his redeemed children, in
the kingdom of his Father.

One of Ward's most earnest desires was, that his
mother should be with him on one of the occasions re-
ferred to. He had, therefore, expressed a wish that she
should not be sent for again till the following week ; by
which time, he had indulged the hope that he might be
so free from suffering as to be able to go to church, if
not actually recovering his usual health and strength.

Instead of the fulfillment of this hope, we had now
to send for that bereaved mother, that she might see the
face of her first-born once more before it should be hidden from her forever, until the dawn of the resurrec-
tion-day.

But when she came in the dead of the night on
Tuesday, her heart failed her ; and she felt that she
could not endure to look on that face in death, which
had been her life, and pride, and joy. She seemed
overwhelmed with grief. "And yet," she said, "the

bitterness of death was past when I parted with him at the hospital, seven weeks ago. I knew I should never see him again on earth; he was too ready for heaven. And that warm, beautiful smile in his eyes, as he looked after me, I would rather keep to remember than the cold sight of his face in death."—*Miss Marsh.*

THE hand of the reaper
  Takes the ears that are hoary,
But the voice of the weeper
  Wails manhood in glory.
The autumn winds rushing
  Waft the leaves that are serest,
But our flower was in flushing
  When blighting was nearest.

"BRING ME UP SAMUEL."

"BRING me Samuel," cries he who disregarded Samuel while living. And so it often is. The father and mother who taught you the right ways of the Lord, have been met by your contempt and disobedience. But the days are coming when their meek, remonstrant faces shall flit before you, and when you will long to bring them back, that you might learn from them the secret of their happiness and their power. Beside the tomb of your parents you will be ready to long that you could bring them again, that you might bewail your undutiful neglects, and make even this tardy reparation for the dishonor you have done them. For what blessing of your better days is not associated with their per-

sons so closely that you can not think of youthful joys
without thinking of THEM?   And what instructions can
ever compare with those which were the first, the sim-
plest, and the most loving?   If you had the power of
raising the dead, in your hour of woe, your language
would not be, " Bring me up the ministers of my mirth
—my comrades in wassail and the dance—my flatterers,
my deceivers, the partners of my avarice and my pomp
— the serpents that twined about me and stung me ;"
but, " Bring me up the ' old man' whose gray hairs I
brought down with sorrow to the grave!   Bring me
up HER who loved me, even in my waywardness ; who
tried to counsel me, even when I would not hearken ;
who comforted me in illness, and who died breathing
prayers in my behalf."—*J. W. Alexander.*

We missed that happiness we might have found ;
A friend is gone, perhaps a son's best friend,
A father, whose authority, in show
When most severe, and mustering all its force,
Was but the graver countenance of love.
We loved, but not enough, the gentle hand
That reared us.   At a thoughtless age, allured
By every gilded folly, we renounced
His sheltering side, and willfully forewent
That converse which we now in vain regret.
How gladly would the man recall to life
The boy's neglected sire !   A mother, too,
That softer friend, perhaps more gladly still,
Might he demand them at the gates of death ;
But not to understand a treasure's worth,
Till time has stolen away the slighted good,
Is cause of half the poverty we feel,
And makes the world the wilderness it is.

DEATH.

ALAS! he is the unsparing invader of every household; all our precautions, all our wisest expedients, in vain are employed to disarm him of his power, and arrest his advancing footsteps. He reigns on earth with a terrible ubiquity. He comes in the hour least expected — often just when the fondest visions of earthly joy are being realized. Do we think of it — we who may be living all careless and thoughtless, lulled by the dream of prosperity, presuming on our present cloudless horizon — that each moment, with sleepless vigilance, the stealthy foe is creeping nearer and nearer? that the smooth current is gliding slowly but surely onward and still onward toward the brink of the cataract, where all at once the irrevocable leap will and must be taken? Reader, perchance you can even now tell the tale! You may be marking the vacant seat at your table, missing the accents of some well-known voice, or the sound of some well-remembered footfall; a beaming eye in your daily walk may be GONE forth forever of time.—*Macduff.*

Oh! how one blow can metamorphose life;
Transmute into the saddest what was once
The happiest home, and open bleeding wounds
Which Heaven alone can medicate!
     Where is the voice whose music
Was more to me than all the world beside?
The noon-day sun his dazzling lustre pours;
Those winged choristers now tune their notes
Around that grave! The bursting loveliness
Of the incipient year, seems but to mock
The desolated spirit which is destined
To know no springtime.

THE DYING INFANT.

You must think, too, of the little sister who is waiting for us in that new home. You remember how I
treasured up the little soft, brown ringlet; her little
well-worn shoes and broken toys ; but you never knew
how much I grieved and mourned for her. Her death
made a life-long impression on me. Twenty-six years
have passed since she has been in glory, but still she is
loved and LONGED for. The anniversary of her death
has just passed, and I have been recalling my feelings
as I went through those deep waters, as she passed
along the dark valley of death. Oh! how agonized I
felt as I stood by her crib and witnessed sufferings we
were unable to alleviate ! How I gazed at her altered
and emaciated, but still beautiful face, as her eyes would
eagerly follow us as we crossed the room to give her
the tea-spoonful of iced water ; and then, as soon as she
would swallow it, the parched lips would beg for the
" drink, drink." Even now my agony comes back, and
I weep as I recall her sufferings. And then the next
day ! the THIRST was gone, but oh! the expression of
that dying face; eternity—HEAVEN can hardly make me
forget it! the infantine expression was all gone, and
she gazed into my face with a woman's intelligence. I
was awed. Sorrow was swallowed up in the feeling
that Death was there — the king of terrors struggling
triumphantly with my child. The little creature would
fix her eyes on me so anxiously, as if she wished to
communicate something, and then she would look up
to the ceiling, as if she was listening earnestly. I drew

my dear old friend, Mrs. M——, down to me, and said:
" Oh! how dreadful, how dreadful! What makes her
look so? she seems to be listening with so much interest to something." Mrs. M—— replied: " Yes;
how earnestly and intelligently she looks up — perhaps
angels are making known to her the plan of salvation
before she meets her Saviour."—*A. N.*

SHE is not dead—the child of our affection—
  But gone into that school
Where she no longer needs our poor protection,
  But Christ himself doth rule.

Day after day we think what she is doing
  In those bright realms of air;
Year after year, her tender steps pursuing,
  Behold her grown more fair.

Thus do we walk with her and keep unbroken
  The bond which nature gives;
Thinking that our remembrance, though unspoken,
  May reach her where she lives.

DEATH OF CHILDREN.

THERE is something exceedingly mysterious in the
early death of the finest children. Nevertheless, we
may not charge God foolishly. You know well how
sometimes you would take the little object of its fond
regard out of the hand and eager grasp of your dear
little child, not in stern severity, but to allure its greater
willingness to come to yourself. God dealeth with us

as with children ; he snatches from us, it may be in the bud, the finest specimens of our nature, around which the fondness and the hope of our hearts cling, not because he would cast us off, but that he may the more effectually win our thoughts and our hearts to himself here, and the more easily reconcile us hereafter to be likewise ever with the Lord.

Tell Mrs. B—— to dry up her tears ; she gave her little darling to the Lord, and where would a mother's heart wish him to be, but just where he is, far better ? I often think of that most wondrous saying of Christ's : " Go thy way, thy son liveth." Ay ! the babe that slept so sweetly in his mother's arms, sleeps in Jesus— he sleeps only ; and " they shall be mine, saith the Lord, in the day when I make up my jewels."

Have the goodness to tell Mrs. B——, from me, not to feel herself less a joyful mother of children, that the Lord had need of her darling George, and wished him nearer himself. It is but a little while, when this thin veil of clouds, hanging its darkness betwixt us and that region of brightness, shall break away, and our God shall put to shame our weeping, giving us back our lost clad in heaven's own garb and beaming in all the light and health of that happiness and glory in which they have been kept and nursed and nourished. " Them that sleep in Jesus will God bring with him." —*John Jamieson.*

> God bless thee ! my beloved child,
> As thou hast blessed me ;
> Faith, peace, and love beyond the grave
> Have been thy gifts to me.

Remembering thee, I look above;
Remembering, wait below,
Trusting with humble confidence,
And patient in my woe.
To me thy early grave appears
An altar for my prayers and tears.

THE SUPREME LOVE OF THE CREATURE— IDOLATRY.

FROM all idolatry our God will cleanse us, and from all our idols Christ will wean us. We may love the creature, but we must not love the creature more than the Creator. When the Giver is lost sight of and forgotten in the gift, then comes the painful process of weaning! When the heart burns its incense before some human shrine, and the cloud as it ascends veils from the eye the beauty and excellence of Jesus—then comes the painful process of weaning! When the absorbing claims and the engrossing attentions of some loved one are placed in competition and are allowed to clash with the claims of God, and the attentions due from us personally to his cause and truth—then comes the painful process of weaning! When creature-devotion deadens our heart to the Lord, lessens our interest in his cause, congeals our zeal and love and liberality, detaches us from the public means of grace, withdraws from the closet, and from the Bible, and from the communion of saints, thus superinducing leanness of soul, and robbing God of his glory—then comes the painful

process of weaning! Christ will be the first in our
affections. God will be supreme in our service — and
his kingdom and righteousness must take precedence
of all other things. In this light, read the present
mournful page in your history. The noble oak that
stood so firm and stately at thy side, is fallen; the tender and beautiful vine that wound itself about thee, is
smitten; the delicate flower that lay upon thy bosom
is withered; the olive-plants that clustered around thy
table are removed, and " the strong staff is broken and
the beautiful rod," not because thy God did not love
thee, but because he desired thine heart.— *Octavius
Winslow.*

EARTHLY love
Must be subordinate to that of heaven,
Or else must die! The earthly gourd
It is permitted thee to cherish fondly,
But not *too* fondly—to be glad for it,
But warning accents from the blighted booth
Of Nineveh, forbid thee to be glad
" Exceedingly."
How oft in one brief day, the canker-worm
Has thus performed its work, and round the bower
Of earthly bliss lie strewn the sad rebukes
Of overweening love—the withered blossoms
Cherished too fondly!

SUFFERING AND SERVING.

THERE is a suffering as well as a doing service. As the exercise of the passive graces is the most difficult, so perhaps it is the most impressive. We peculiarly glorify God in the fires. We are witnesses for him, and testify to the excellency of the principles, and to the power of the resources of the religion we profess. We know that his religion can support us when every other dependence fails, and his comfort cheer us when all other springs of comfort are dried up. When, by accident or sickness, we are led in from active scenes, we fear we are going to possess months of vanity, whilst perhaps we are entering some of the most useful parts of our life. If we endure as Christians, the spirit of glory and of God resteth upon us; and by our patience, submission, peace, and joy, some around us are instructed, some convinced, some encouraged — while perhaps superior beings are excited to glorify God in us, for we are a spectacle to angels as well as men.—*Rev. William Jay.*

ONCE, when young Hope's fresh morning dew
Lay sparkling on my breast,
My bounding heart thought but to DO,
To WORK, at Heaven's behest. My PAINS
Come at the same behest!

All fearfully, all tearfully,
Alone and sorrowing,
My dim eye lifted to the sky,
Fast to the cross I cling—O Christ!
To thy dear cross I cling.

A LITTLE WHILE.

"Yet a little while, and He that shall come will come, and will not tarry."—Hebrews 10 : 37.

"A little while!" and then sorrow, suffering, tears, death, sin, will be known no more! Let me compose myself to sleep, or rest my aching head on its pillow, with the joyous thought: "Soon to be with Christ, and that forever and ever."—*Soldier's Text-Book.*

> Oh! for the peace which floweth as a river,
>   Making life's desert places bloom and smile!
> Oh! for the faith to grasp heaven's bright "forever,"
>   Amid the shadows of this "little while!"
> "A little while" for patient vigil keeping,
>   To face the storm, to wrestle with the strong;
> "A little while" to sow the seeds with weeping,
>   Then bind the sheaves and sing the harvest-song.

"PEACE, BE STILL."

"They said one to another: 'What manner of man is this, that even the winds and the sea obey him?'" Their Lord rose higher than ever in their estimation. In the future manifold sacred memories of that wondrous ministry, how the combined remembrance of the WEARY man and the ALMIGHTY God would brace them for their great fight of afflictions! That "Peace, be still," has been a motto and a watchword, which those

howling winds of Gennesaret have wafted from age to
age, and from clime to clime, sustaining faith in sinking
hearts, and producing in many a storm-swept bosom a
" great calm."—*Macduff*.

> Oh ! for a faith that will not shrink,
>   Though pressed by every foe ;
> That will not tremble on the brink
>   Of any earthly woe.
>
> That will not murmur or complain,
>   Beneath the chastening rod ;
> But in the hour of grief or pain
>   Will lean upon its God.
>
> A faith that shines more bright and clear
>   When tempests rage without ;
> That when in danger knows no fear,
>   In darkness feels no doubt.

THE CREATURE AND THE CREATOR.

THENCE it is, because God alone is our last end, that
he alone never FAILS us. All else fails us but he. Alas!
how often is life but a succession of worn-out friendships ? Youth passes with its romance, and crowds
whom we loved have drifted away from us. They
have not been unfaithful to us, nor we to them. We
have both but obeyed a law of life, and have exempli-
fied a world-wide experience. The pressure of life
has parted us. Then comes middle life, the grand
season of cruel misunderstandings, as if reason were

wantoning in its maturity, and by suspicions and cir-
cumventions and constructions were putting to death
our affections. All we love and lean upon fails us. We
pass through a succession of acquaintanceships; we tire
out numberless friendships; we use up the kindness of
kindred; we drain to the dregs the confidence of our
fellow-laborers; and there is a point beyond which we
must not trespass on the forbearance of our neighbors.
And so we drift on into the solitary havens of old age,
to weary by our numberless wants the fidelity which
deems it a religion to minister to our decay. And then
we see that God has outlived and outlasted all: the
Friend who was never doubtful; the Partner who never
suspected; the Acquaintance who loved us better at
least it seemed so—the more evil he knew of us; the
Fellow-laborer who did our work for us as well as his
own; and the Neighbor who thought he had never done
enough for us; the one Love that, unlike all created
loves, was never cruel, exacting, precipitate, or over-
bearing. *He has had patience with us, has believed in
us, and has stood by us.* What should we have done
if we had not had him? All men have been liars; even
those who seemed saints broke down when our imper-
fections leaned on them, and wounded us, and the
wound was poisoned; but He has been faithful and true.
On this account alone, he is to us what neither kins-
man, friend, nor fellow-laborer can be.—*Faber.*

EARTH's light all faded, and shaken all trust,

Steals now a soothing voice on her rapt ear—
" Lean on ME, daughter, and be of good cheer;

Render not worship, that worketh such woe—
Thy nature's deep cravings God only can know."

Hushed is the tempest, the eyes glance above,
Yearns the lone heart to the Father of Love;
Pleading in low tones for heaven's calm rest—
"Disappointed in ALL, take me home to thy breast."

"Trust in *Me*, daughter, and toil on awhile,
Guided and warmed by the light of my smile;
A mission of love, to the stricken and lone,
Be thine to fulfill, child, forever mine own."

Humbly then turns she her duties to meet,
Fainting, yet eager her task to complete;
Earth's shadows around her, but light in her soul—
The Father—Friend—beckons her on to the goal.

DEATH OF A DAUGHTER.

SHE who was the sweet singer of my little Israel is no
more. The child whose sense of beauty made her the
swiftest herald to me of every fair discovery and new
household joy, will never greet me again with her sur-
prises of gladness. She who, leaning upon my arm as
we walked, silently conveyed to me such a sense of
evenness, firmness, dignity; she whose childlike love
was turning into the womanly affection for a father; she
who was complete in herself, as every good child is, not
suggesting to your thoughts what you would have a
child be, but filling out the orb of your ideal beauty,
still partly in outline; her seat, her place at the table, at
prayers, at the piano, at church; the sight of her going

out and coming in; her tones of speech, her helpful
spirit and hands, and all the unfinished creations of her
skill; every thing that made her that which the growing
associations with her name had built up in our hearts—
all is gone, for this life. It is removed like a tree; it is
departed like a shepherd's tent.

And all this, too, is saved. It survives, or I would
not, I could not, write thus. There comes to my sorrowing heart some such message as the sons of Jacob
brought to their father, when they said: "Joseph is
yet alive, and he is governor over all the land of Egypt."

Jesus of Nazareth has been in my dwelling, and has
done a great work of healing. He has saved my child;
saved her to be a happy spirit; forever saved her for
himself, to employ her powers of mind and heart in his
blissful service. He has saved her for me through all
eternity. She will be my sweet singer again; she will
have in store for me all the wonderful discoveries which
her intense love of beauty will have made her treasure
up, to impart, when the child becomes, as it were, parent
for a little while, to the soul of the parent, in heaven,
new-born.—*Nehemiah Adams.*

WHEN the shaded pilgrim-land
    Fades before my closing eye,
Then revealed on either hand,
    Heaven's own scenery shall lie;
Then the veil of flesh shall fall,
    Now concealing, darkening all.

Heavenly landscapes, calmly bright,
    Life's pure river murmuring low,

Forms of loveliness and light,
  Lost to earth long time ago;
Yes, mine own, lamented long,
  Shine amid the angel throng.

When upon my wearied ear
  Earth's last echoes faintly die,
Then shall angels' harps draw near—
  All the chorus of the sky;
Long-hushed voices blend again,
  Sweetly in that welcome strain.

INTIMACIES OF EARTH RENEWED IN GLORY.

Our Bibles, in manifold direct as well as indirect passages, foster the inspiriting hope, that the hallowed intimacies of earth will be renewed and perpetuated in glory. The thought of the loved and lost — now the loved and glorified—being "the *loved* and *known* again;" does not this tinge our every anticipation of heaven with a golden hue, and form a new and holy link binding us to the throne of God?

Our blessed Lord himself, alike by his discourses and his example, has strengthened our belief in the future reünion and recognition of saints. He speaks of "Abraham, Isaac, and Jacob," as distinct persons in the kingdom of heaven. He speaks of "the beggar"—the identical person laid on earth at "the rich man's gate"—now "in Abraham's bosom." When he comforted the hearts of the bereaved sisters of Bethany, his consolatory announcement was not, "Lazarus shall rise," but

"Your BROTHER shall rise again." Affection was to be restored at the great day; the brother of the earthly was to be known and welcomed as brother in the heavenly home.

On Mount Tabor, Moses and Elias came down, in form and feature the same as they were when they dwelt in their earthly tabernacles.

Yes; I fondly cling to the hope—the BELIEF—that in heaven there will be joyful reunions and recognitions. The grave will not be permitted to efface the memorials of the past, and destroy our personal identity. The resurrection-body will wear its old smiles of love and tenderness. "THEM ALSO THAT SLEEP IN JESUS (literally, LAID TO SLEEP BY JESUS) WILL GOD BRING WITH HIM."—*Grapes of Eschol.*

When no shadow shall bewilder,
When life's vain parade is o'er,
When the sleep of sin is broken,
And the dreamer dreams no more,
When the bond is never severed—
Partings, claspings, sobs, and moans,
Midnight-waking, twilight-weeping,
Heavy noontide—all are done;
When the child has found its mother,
When the mother finds the child;
When dear families are gathered,
That were scattered on the wild;
Brother, we shall meet and rest,
'Mid the holy and the blest.

DISCIPLINE.

FAITH considers LOVE as the motive on God's part of all afflictions. They not only come on those whom God loves, but BECAUSE he loves them. They are love-tokens as much as any thing else that comes from the hand of love. The father chastens his son in love—gives him medicine in love—denies him some things he asks for in love. It is the severity of love, I admit, but still it IS love, and a contrary line of conduct would not be love. But often it requires strong faith to believe this. "What! this love, to wither my gourd, and scorch my head by the sun, and beat upon me by his fierce hot blast? This love, to shatter my cisterns, and spill their water upon the ground? This love, to frustrate my schemes and disappoint my hopes, and strip me of my comforts? This love, to fill my eyes with tears and my bosom with sighs?" "Yes," replies God, "As many as I love, I rebuke and chasten." "Enough," says the Christian, "I believe it; and my soul is even as a weaned child."—*J. A. James.*

TREMBLE not, though darkly gather
Clouds and tempests o'er thy sky;
Still believe thy Heavenly Father
Loves thee best when storms are nigh.

Love divine has seen and counted
Every tear it caused to fall;
And the storm which love appointed
Was its choicest gift of all.

CHRIST PRECIOUS.

THE truth is that we never feel Christ to be a reality until we feel him to be a NECESSITY. Therefore, God makes us feel that necessity. He tries us here, and he tries us there. He chastises on this side, and he chastises on that side. He probes us by the disclosure of one sin, and another, and a third, which have lain rankling in our deceived hearts. He removes, one after another, the objects in which we have been seeking the repose of idolatrous affection. He afflicts us in ways which we have not anticipated. He sends upon us the chastisements which he knows we shall feel most sensitively. He pursues us when we would fain flee from his hand; and, if need be, he shakes to pieces the whole framework of our plans of life, by which we have been struggling to build together the service of God and the service of self; till, at last, he makes us feel that Christ is all that is left to us.—*Austin Phelps.*

In the dark winter of affliction's hour,
 When summer friends and pleasures haste away,
And the wrecked heart perceives how frail each power
 It made a refuge and believed a stay;
When man all vain and weak is seen to be—
There's none like THEE, O Lord! there's none like thee!

When the world's sorrow working only death,
 And the world's comfort, caustic to the wound,
Make the wrung spirit loathe life's daily breath,
 As jarring music from a harp untuned;
While yet it dare not from the discord flee—
There's none like *thee*, O Lord! there's none like thee!

DEATH OF AN AGED CHRISTIAN.

THE aged disciple of Jesus — why should we wish to detain him? His work is done. Why desire to hold him back from the grave? It is through the gate and grave of death that he passes to his inheritance above. Why be inconsolable at his departure? He is not lost, neither is the light of his mind or heart extinguished. Why mourn as those who have no hope, beside his tombstone? He shall not lie there long. He is planted there in the likeness of Christ's death, that he may rise with Christ to the resurrection of eternal life. Not many days shall roll over you ere you and they shall all rise again; "they that have done good to the resurrection of life, and they that have done evil to the resurrection of damnation." Rejoice, rather, when one you love, who is full of days and full of grace, sets like a sun behind the horizon of life. Rejoice, for he shall rise again; and when that morning of the resurrection dawns, it will usher in a day that has no clouds, a day that has no sunset, and a day that is followed by no night of sorrow or of death.— *W. B. Stevens.*

THEN rose another hoary man and said,
   In faltering accents to that weeping train:
Why mourn ye that our aged friend is dead?
   Ye are not sad to see the gathered grain,
Nor, when their mellow fruit the orchards cast,
Nor when the yellow woods shake down the ripened **mast.**

Why weep ye then for him, who having won
   The bound of man's appointed years, at last,

Life's blessings all enjoyed, life's labors done,
　　Serenely to his final rest has passed;
While the soft memory of his virtues yet
Lingers, like twilight hues, when the bright sun is set?

And I am glad that he has lived thus long,
　　And glad that he has gone to his reward;
Nor can I deem that nature did him wrong,
　　Softly to disengage the vital cord;
For when his hand grew palsied, and his eye　　·
Dark with the mists of age, it was his time to die.

LEADING THE BLIND.

WE should naturally expect Christians as being enlightened, and knowing more of God and the ways of
God than irreligious men; we should naturally expect
THEM to have more correct expectations of God's treatment of them. But they are slow to learn; they are
often disappointed; their anticipations are no foreshad-
owing of God's treatment of them. Their comforts,
their prosperity and strength seldom come to them in
the way of their anticipations; yea, VERY seldom or
never. The allotments of Divine providence which affect them most are such as they little expected. Some
of the evils they have suffered were evils which they
struggled hard and prayed hard to escape. But God
would not let them off. His unseen hand pushed them
steadily on right into the cloud and the calamity which
they most dreaded. Out of these calamities, out of
these griefs and shocks and shiftings, which they deem-

ed curses, God gave them the most signal of their benefits, teaching them best to know him, to trust him, and distrust themselves. " He led them in a way they knew not."

There are some, yea, there are many with whom God hath dealt more favorably than their fondest expectations. His smiles, his prosperities have attended them all along, and all along their hearts have been overwhelmed, and their souls become more humble and holy, by a sense of the goodness and mercy and bounty of God. They never expected such days of sunshine. They expected storms. They knew — have always known — that no fidelity in them gave them any claim or ground to expect favors; and now, when they contemplate them, and look back, and try to number up their mercies, love, gratitude, faith fill their minds, and fill them most of all because God's outward benefits have not led them to forget him. There are some such; yea, (let us do religion justice,) there are many such. And just like the others, they have been led in paths they never anticipated. Indeed, I believe it is almost universal with Christians, when they remember divine providences which have affected them, and especially when they remember how they have been spiritually dealt with, I believe it is almost universal with them to wonder and praise and adore God that he has led them in a way they knew not—his way, not their own.

If God is leading us on toward heaven, he will COMPEL us to trust him. We are blind; we need him to lead us. Often he confounds our counsels, defeats our

purposes, disappoints our hopes, and drives us into
difficulties; yea, sometimes into despair, just to bring
us to that sweeping and sweet faith which puts every
thing into his hands, and trusts him in the dark. By
such a faith DARKNESS BECOMES LIGHT. It makes us
know God better, and Christ better, and grace better.
Never point out a way for yourself. Take God's way.
—*J. S. Spencer.*

SEND kindly light amid the encircling gloom,
　　And lead me on;
The night is dark, and I am far from home,
　　Lead thou me on!
Keep thou my feet: I do not ask to see
The distant scene; ONE STEP enough for me.

I was not ever thus, nor prayed that thou
　　Shouldst lead me on;
I loved to choose and see my path; but now
　　Lead thou me on!
I loved day's dazzling light, and spite of fears
Pride ruled my will: remember not past years!

DEATH OF A HUSBAND.

OH! how earnestly I wished to go with him! I was
for the time insensible to my own loss; my soul pursued
him into the invisible world; and for the time I cordially rejoiced with the Spirit. I thought I saw the angel-
band ready to receive him, among whom stood my dear
mother, the first to bid him welcome to the regions of
bliss. I was desired to leave the room, which I did,

saying: " My doctor is gone. I have accompanied him to the gates of heaven; he is safely landed." I went into the parlor. Some friends came in to see me. My composure they could not account for. Our sincere and tender regard for each other was too well known to allow them to impute it to indifference. In the evening I returned to the bed-chamber, to take a last farewell of the dear remains. The countenance was so very pleasant, I thought there was even something heavenly, and couldn't help saying: " You smile upon me, my love. Surely the delightful prospect, opening on the departing soul, left that benign smile on its companion, the body." I thought I could have stood and gazed for-ever; but, for fear of relapsing into immoderate grief, I withdrew after a parting embrace. I went to bed purely to get alone, for I had little expectation of sleep. But I was mistaken; nature was fairly overcome with watching and fatigue. I dropped asleep, and for a few hours forgot my woes; but, oh! the pangs I felt on first awaking! I could not for some time believe it true that I was, indeed, a widow, and that I had lost my heart's treasure; my all I held dear on earth. It was long before day. I was in no danger of closing my eyes again, for I was at that time abandoned to despair, till recollection and the same considerations which at first supported me brought me a little to my-self. I considered that I wept for one that wept no more; that all my fears for his eternal happiness were now over, and he beyond the reach of being lost; neither was he lost to me, but added to my heavenly treasure more securely mine than ever.—*Isabella Graham.*

So, hand in hand, we trod the wild,
  My angel-love and I,
His lifted wing all quivering
  With tokens from the sky.
Strange my dull thought could not divine
  'Twas lifted but to fly.

Again down life's dim labyrinth
  I grope my way alone,
While wildly through the midnight sky
  Black hurrying clouds are blown,
And thickly in my tangled path
  The sharp, bare thorns are sown.

Yet firm my foot, for well I know
  The goal can not be far,
And ever through the rifted clouds
  Shines out one steady star;
For when my guide went up, he left
  The pearly gates ajar.

EXTRACT FROM A FUNERAL SERMON.

SHE has "fallen asleep," as the child, weary of weeping, sometimes turns in the mother's arms and rests. And parental solicitude, retrospective of a thousand particulars which none but a father or a mother can comprehend, will acquiesce in such relief and escape from trial. We speak so often, my brethren, of the domestic relations, that we are apt to forget how profound are the sentiments to which they give rise. Some there are who treat as exaggerations much that

is said and written concerning the warmth of attach-
ment between parent and child, brother and sister,
friend and friend. I profess myself to be of the
mind of those who believe that the affection of a parent, purified by religion, may equal the highest reaches
of romance and poetry. But there are chords which
the hand even of sympathetic friendship may jar too
roughly. The words of human speech can not tell
how great, how tender the deposit of treasured love
which lies in those cerements. Beloved friends, not
only resign yourselves, but hush all wishes! God has
sweetly interposed, and his touch is love. She whom
you cherished, and embraced all the more yearningly,
if at any time she speeded from the howling tempest to
nestle in your bosom, longed for the infinite solace, and
could be content with no earthly covert; wandering in
quest of peace, she found no rest for the sole of her
foot, till she burst from that fainting body. She is
with the Lord of peace. There the weary are at rest.
Jesus, whom she sought and loved, has at length, ear-
lier than she or we expected, met her with the kiss of
peace. He has stooped to wipe the moisture of weariness and anguish from her marble brow. He has taken
her in his arms, out of the last fatal swooning. He has
said to her, "Mary," and she has answered: "Rabbo-
ni!"—*J. W. Alexander.*

Mother! why grieve for me?
I've reached my heavenly home;
Your wearied pilgrim rests at last,
I'm sheltered from the storm.

Life's hard, rough road is trod,
    I've crossed the stormy sea;
Those storms, they brought me to my God;
    You should rejoice with me.

Why do you mourn for me?
    I have no trouble here;
Each suffocating sob is stilled,
    Dried is each burning tear!

Joy now lights up my brow,
    Peace has returned to me;
The future can not cheat me now,
    E'en the PAST seems bright to me!

DEATH WELCOME.

SOMETIMES in pacing the shore of that great ocean which you are so soon to cross, solemn thoughts have arisen: "Why this clinging to mortality? why this love of life, this fear of death? Can I belong to Christ, and yet so deprecate departing to be with him?" But if you are really his, he will arrange it all most excellently. The believer will tarry till he can say: "Now, Lord, lettest thou thy servant depart in peace." And this the Lord usually effects by loosening that chain which held him to this life, or by presenting such a strong attraction that the chain is broken unawares. The summer before good old Professor Wodrow died, Principal Sterling's lady came in to see him, and he said to her: "Mrs. Sterling, do you know the place in the new kirkyard that is to be my grave?"

She answered, she did. "Then," says he, "the day is good, and I'll go through the Principal's garden into it, and take a look at it. Accordingly they went, and when they came to the place, as near as she could guess, she pointed it out to him, next to Principal Dunlop and her own son and only child. He looked at it, and lay down upon the grass, and stretched himself most cheerfully on the place, and said : "Oh! how satisfying it would be to me to lay down this carcass of mine in this place, and be delivered from my prison! But it will come in the Lord's time!" But, although for more than forty years this cheerful Christian had never one day doubted his heavenly Father's love, it was not till his own dear children had gone before, and till manifold infirmities made the flesh a burden, that he felt thus eager to put off the tabernacle.—*Hamilton.*

FATHER ! into thy loving hands
  My feeble spirit I commit,
While wandering in these border-lands,
  Until thy voice shall summon it.

These border-lands are calm and still,
  And solemn are their silent shades ;
And my heart welcomes them, until
  The light of life's long evening fades.

They say the waves are dark and deep,
  That faith hath perished in the river ;
They speak of death with fear, and weep :
  Shall my soul perish ? Never, never!

And I will calmly watch and pray
Until I hear my Saviour's voice
Calling my happy soul away
To see his glory, and rejoice.

THE PAST.

IT is wisest, when we can do it, to put away the past altogether; we have done with it in the way of action, we can not improve it by way of thought. We have a future, at least we have a present, where effort need not be spent in vain. But it is sexton's work to linger moralizing perpetually amongst the graves. If we have strength, close we that inevitable gate, and go forth amongst the striving throng to live and labor, to wait and pray.—*Holme Lee.*

Not enjoyment and not sorrow
Is our destined end and way;
But to act, that each to-morrow
Finds us further than to-day.

Let us then be up and doing,
With a heart for any fate;
Still achieving, still pursuing,
Learn to *labor* and to wait.

SORROW FOR THE DEAD.

SORROW for the dead is the only sorrow from which we refuse to be divorced. Every other wound we seek to heal; every other affliction to forget; but this wound we consider a duty to keep open; this affliction we cherish and brood over in solitude. Where is the mother that would willingly forget the infant that perished like a blossom from her arms, though every

recollection is a pang? Where is the child that
would willingly forget the most tender of parents,
though to remember be but to lament? Who in
the hour of agony would forget the friend over
whom he mourns? Who, even when the tomb is
closing upon the remains of her he most loved, and
he feels his heart, as it were, crushed in the closing
of its portal, would accept consolation that was to be
bought by forgetfulness? No; the love which survives the tomb is one of the noblest attributes of the
soul. It has its woes; it has likewise its delights; and
when the overwhelming burst of grief is calmed into
the gentle tear of recollection, when the sudden anguish and the convulsive agony over the ruins of all
that we most loved, is softened away into pensive meditation on all that it was in the days of its loveliness,
who would root out such a sorrow from the heart?
Though it may sometimes throw a passing cloud over
the bright hour of gayety, or spread a deeper sadness
over the hour of gloom, yet who would exchange it for
the song of pleasure or the burst of revelry? No;
there is a voice from the tomb sweeter than song;
there is a recollection of the dead to which we turn
even from the charms of the living. — *Washington
Irving.*

TILL my heart dies, it dies away
In yearnings for what might not stay;
For love which ne'er deceived my trust,
For all which went with "dust to dust."

We miss them when the board is spread;
We miss them when the prayer is said;

Upon our dreams their dying eyes
In still and mournful fondness rise.

Holy ye were, and good and true!
No change can cloud my thoughts of you;
Guide me, like you, to live and die,
And reach my Father's house on high!"

THE SEA A CEMETERY.

WHEN it thunders and lightens, I often think how secure the little sleeper is, and when the rain comes down on that peaceful grave, my heart betakes itself to calm thoughts, because the precious dust feels no tempests, wakes at no alarm. The loss of that passenger-ship with four hundred souls on board made me think, what a cemetery is the sea! None are thought of, loved, and mourned over more than they who find their sepulture there. It is soothing to have the dust of a child or friend in a sure, safe grave, when you meet with those whose loved ones are lost in the great waters. But He who is the "Resurrection and the Life" has his eye upon them. The Lord buried them, and no man knoweth of their sepulchres.—*Nehemiah Adams.*

. . . . . . SHE lay a thing for earth's embrace,
To cover with spring wreaths. For earth's? The wave
That gives the bier no flowers, makes moan above her grave!
. . . . the voice of prayer,
And then the plash in the deep waters! Thy bed
Is under the restless wave, my Elinor—

Thy lullaby, the ocean's moan ; and never more,
Loved as thou wert, may human tear be shed
Above thy rest ! No mark the proud seas keep
To show where he that wept may pause again to weep.

So the depths took thee ! Oh ! the sullen sense
Of desolation in that hour compressed !
Dust going down, a speck amidst the immense
And gloomy waters, leaving on their breast
No trace of the heart's idol ! Blest are they
That earth to earth intrust, for they may know
And tend the dwelling where the slumberer's clay
Shall rise at last, and bid the young flowers bloom,
That waft a breath of hope around the tomb—
And kneel upon that precious turf to pray !

MUCKLE KATE.

NOT only was she satisfied in regard to her eternal safety, but she had attained that enviable point at which assurance had become so sure that she ceased to think of self, and so wholly was she absorbed in the glory of her Redeemer, that even to herself she was nothing—Christ was all in all. The glory of Christ was her all-engrossing motive. The inexpressible joy that was vouchsafed her served but to quicken her departing soul to more rapturous commendations to others of that Saviour whom she had found ; and when at length the welcome summons came, and she stood upon the threshold of eternal glory, ere yet the gate had fully closed upon her ransomed spirit, the faltering tongue was heard to exclaim, as its farewell effort in Christ's

behalf: "Tell, tell to others that I have found him."
Lay the emphasis upon the "I," and behold the world
of meaning condensed into those dying words. Com-
press into that "I" those ninety years of sin, and you
catch its full force. "Tell them that I, the worst of
sinners, the drunkard, the profligate, the Sabbath-
breaker, the thief, the blasphemer, the liar, the scoffer,
the infidel—tell them that I, a living embodiment of
every sin, even I have found a Saviour's person, even I
have known a Saviour's love."—*T. M. Fraser.*

LOOKING to Jesus with a steadfast eye,
    Clad in his righteousness, my robe divine,
Come! for thy boasted terrors I defy,
    Poor, harmless, shadowy phantom! He is mine;
My life is bound in his whose living word
Cries that the dead are blest when dying in the Lord.

I see him shining on his throne of light,
    The Lamb that hath been slain, and slain for *me;*
The King of glory! of all power and might,
    The Lord and God, by whose most high decree
The vile, the guilty, trusting in his name,
A dying wretch like me, eternal life may claim.

LOSS OF A WIFE.

I HAVE returned HITHER, but have an utter repug-
nance to say returned HOME—that name is applicable
no longer. You may be sure I am grateful for your
kind sympathy and suggestions of consolation, not the
less so for its being too true that there is a weight on

the heart which the most friendly human hand can not remove. The melancholy fact is, that my beloved, inestimable companion has left me. It comes upon me in evidence how varied and sad! and yet for a moment sometimes I feel as if I could not realize it as true. There is something that seems to say: Can it be that I shall see her no more, that I shall still, one day after another, find she is not here; that her affectionate voice and look will never accost me; the kind grasp of her hand never more be felt; that when I would be glad to consult her, make an observation to her, address to her some expression of love, call her "my dear wife," as I have done so many thousand times, it will be in vain— she is not here? I have not suffered, nor expect to feel any overwhelming emotions, any violent excesses of grief. What I expect to feel is a long repetition of pensive monitions of my irreparable loss; that the painful truth will speak itself to me again, and still again in long succession, often in solitary reflection, (in which I feel the most,) and often as objects come in my sight, or circumstances arise which have some association with her who is gone.—*John Foster.*

SLEEP on, my love, in thy cold bed,
Never to be disquieted!
My last good night! thou wilt not wake
Till I thy fate shall overtake;
Till age, or grief, or sickness must
Marry my body to that dust
It so much loves, and fill the room
My heart keeps empty in thy tomb.
Stay for me there. I will not fail

To meet thee in that hollow vale ;
And think not much of my delay—
I am already on the way,
And follow thee with all the speed
Desire can make or sorrow breed ;
Each minute is a short degree,
And every hour a step toward thee !
But hark ! my pulse like a soft drum
Beats my approach, tells thee I come ;
And slow howe'er my marches be,
I shall at last lie down by thee !
The thought of this bids me go on
And wait my dissolution.
With hope and comfort, dear, (forgive
The crime,) I am content to live
Divided, with but half a heart,
Till we shall meet and never part.

I AM SATISFIED.

YES, I am satisfied, I am comforted. And if one of
the many involuntary tears I have shed could recall her
to life, to health, to an assemblage of all that this world
could contribute to her happiness, I would struggle hard
to suppress it. Now my largest desires for her are
accomplished. The days of her mourning are ended.
She is landed on that peaceful shore where there are
no storms of trouble. She is forever out of the reach
of sorrow, sin, temptation, and snares. Now she is
before the throne! She sees Him whom, not having
seen, she loved; she drinks of the rivers of pleasure,
which are at his right hand, and shall thirst no more.—
John Newton.

O selfish tears! who would unglorify
The sainted pilgrim? her unruffled bliss
Disturb, and pluck the crown from off her brow
To bring her back to earth? Fallen she has
Asleep in Jesus; basking forever
Beneath the sunshine of Jehovah's smile.
Sorrows all ended, wiped from off her eye
The lingering tear-drop—immortality
Begun.

TRIALS.

THERE is nothing which shows our ignorance so much as our impatience under trouble. We forget that every cross is a *message* from God, and intended to do us good in the end. Trials are intended to make us think, to wean us from the world, to send us to the Bible, to drive us to our knees. Health is a good thing; but sickness is far better, if it leads us to God. Prosperity is a great mercy, but adversity is a greater one if it brings us to Christ. Any thing, any thing is better than living in carelessness and dying in sin.—*Ryle.*

O LORD! I pray thee comfort me
In this my sore and deep distress,
And let my troubled spirit see
The wonders of thy faithfulness.

Shine on this barren ground, that I
Lose not the fruits which should spring up ·
Let me not pass thy mercy by,
Nor miss the sweetness in my cup.

Sweetness there is, I know it, Lord,
And otherwise there can not be;
It is my Father's hand that poured
This mixture in the cup for me.

What is it, Lord? dost thou intend
That patience should take root in me?
Is it thy will my will to bend,
That I more like a child may be?

Is it to raise my heart above
All earthly care and earthly pleasure,
And loose my hands from earthly love,
To fill them full of heavenly treasure?

THE WIDOW'S GOD.

" Let thy widows trust in me."

THE companion of your youth, the friend of your
bosom, the treasure of your heart, the staff of your
riper and the solace of your declining years, is removed;
but since God has done it, it is, it must be WELL. And
who is the object of the widow's trust? "In me,"
says God. None less than himself can meet your case.
He well considers that there is an acuteness in your
sorrow, a depth in your loss, a loneliness and a helplessness in your position, which no one can meet but himself. The first, the best, the fondest, the most pro-
tective of creatures has been torn from your heart, is
smitten down at your side. What other creature could
now be a substitute? A universe of beings could not
fill the void. God in Christ only can. O wonderful

thought! that the divine Being should come and imbosom himself in the bereft and bleeding heart of a human sufferer — that bereft and bleeding heart of YOURS. He is especially the God of the widow. And when he asks your confidence and invites your trust, and bids you lift your weeping eye from the crumbled idol at your feet, and fix it upon himself, he offers you an INFINITE substitute for a finite loss ; thus, as he ever does, giving you infinitely MORE than he took, bestowing a richer and a greater blessing than he removed.

And what are you invited thus to intrust to God ? YOURSELF. God seems now to stand to you in a new relation. He has always been your Father and your Friend. To these he now adds the relation of HUSBAND. You are to flee to him in your helplessness, to resort to him in your loneliness, to confide to him your wants, and to weep your sorrows upon his bosom. You are to trust your CHILDREN into God's hands. He says : "Leave your fatherless children ; I will preserve them alive." "Thou art the helper of the fatherless." "Enter not into the field of the fatherless, for their Redeemer is mighty ; he will plead their cause with thee." He has removed their earthly father that he may adopt them as his own. His promise that he will "preserve them alive," you are warranted to interpret in its best and widest sense. It must be regarded as including, not temporal life only, but also spiritual life. He will preserve your fatherless ones alive temporarily, providing all things necessary for their present existence ; but infinitely more than this, he will, in answer to the prayer of faith, preserve their souls unto eternal life.

Your CONCERNS are to be trusted to God. These, doubtless, press at this moment with peculiar weight upon your mind. They are new and strange. They were once cared for by one in whose judgment you had implicit confidence, whose mind thought for you, whose heart beat for you, whose hands toiled for you, who in all things sought to anticipate every wish, to reciprocate every feeling; whose esteem, and affection, and confidence shed a warm and mellow light over the path of life. These interests once confided to his judgment and control, must now be intrusted to a wiser and more powerful Friend, to Him who is truly and emphatically the widow's God.—*Octavius Winslow.*

NOTHING but perfect trust,
　And love of thy perfect will,
Can raise me out of the dust,
　And bid my fears be still.

Lord, fix my eyes upon thee,
　And fill my heart with thy love;
And keep my soul till the shadows flee,
　And the light breaks forth above!

A DAY OF DISCLOSURES.

BELIEVER, be still! The dealings of thy heavenly Father may seem dark to thee; there may seem now to be no golden fringe, no "bright light in the clouds;" but a day of disclosures is at hand. Take it on trust "a little while." An earthly child takes on trust what

his father tells him. When he reaches maturity, much
that was baffling to his infant comprehension is ex-
plained. Thou art in this world in the nonage of thy
being — eternity is the soul's immortal manhood.
THERE, every dealing will be vindicated. It will lose
all its darkness when bathed in the floods "of the ex-
cellent glory !"— *Words of Jesus.*

"A LITTLE while" to wear the robe of sadness
     And toil with weary step through miry ways;
Then to pour forth the fragrant oil of gladness,
     And clasp the girdle round the robe of praise:
"A little while" 'midst shadow and illusion,
     To strive by faith love's mysteries to spell;
Then read each dark enigma's bright solution,
     And hail sight's verdict: "He doth all things well."

DEATH OF A FATHER.

MARCH 9, Sunday.—Dearest papa's first Sabbath in
" glory everlasting !"   March 13.—Went twice to look
at dearest papa's earthly tabernacle. This corruptible
" SHALL put on incorruption."   March 14.—All that remained of dearest papa buried in the vault at Mick-
leover, till Jesus says: "Come forth !"   It has been a
time of deep and unutterable sorrow, yet mixed with
countless mercies and loving-kindnesses. Indeed, I
often feel far more inclined to rejoice than to weep.
For above an hour after he went, I sat by all that remained to me of him, the greater part of the time being
quite alone, yet not one tear could I shed ! No; I was

absorbed in thoughts of unseen realities, and so marvelously have they taken possession of me since, that I seldom have felt inclined to weep. Buried on a lovely, bright morning, which filled me full of resurrection thoughts! " Lazareth, come forth!" were words I delighted to listen to, the Spirit speaking in the word. It seemed so impossible to think of the tears Jesus shed over the lifeless body of Lazarus, without going on to the omnipotence which said : "Lazarus, come forth!"— *Adelaide Newton.*

> That crumbling framework crumbles but to live !
> Immanuel's blood, which bought the soul, has paid
> The ransom of the body.
>   Repose, then, precious clay !
> Thou art in safer custody than mine,
> The purchase of atoning blood ! What though
> The sods of earth now cover thee, and rage
> The elements around thee—angels watch
> The sleeping dust ; nay more, Omnipotence
> Is the invisible Guardian of the tomb !

DEATH.

Deaths are being died somewhere every moment. But it is not a melancholy thought. Every hour—we feel it most at evening—it is like a balm to our spirits to think of the busy benevolence of death, ending so much pain, crowning so much virtue, swallowing up so much misery, pacifying so much strife, illuminating so much darkness, letting so many exiles into their eternal home

and to the land of their eternal Father! O grave and
pleasant cheer of death! How it softens our hearts, and
without pain kills the spirit of the world within our
hearts! It draws us toward God, filling us with strength
and banishing our fears, and sanctifying us by the pathos
of its sweetness. When we are weary and hemmed in
by life, close and hot and crowded — when we are in
strife and self-dissatisfied—we have only to look out in
our imaginations over wood, and hill, and sunny earth,
and star-lit mountains, and the broad seas, where blue
waters are jeweled with light islands, and rest ourselves
on the sweet thought of the diligent, ubiquitous be-
nignity of death.—*Faber.*

THEY are gathering homeward from every land,
>   One by one,
As their weary feet touch the shining strand,
>   One by one.
Their brows are inclosed in a golden crown,
Their travel-stained garments are all laid down,
And clothed in white raiment they rest on the mead,
Where the Lamb loveth his chosen to lead,
>   One by one.

Before they rest they pass through the strife,
>   One by one;
Through the waters of death they enter life,
>   One by one.
To some are the floods of the river still,
As they ford on their way to the heavenly hill;
To others the waves run fiercely and wild,
Yet all reach the home of the undefiled,
>   One by one.

DEATH OF A DAUGHTER.

THERE is a better world, of which I have thought too little. To that world she has gone, and thither my affections have followed her. This was Heaven's design. I see and feel it as distinctly as if an angel had revealed it. I often imagine that I can see her beckoning me to the happy world to which she has gone. I want only my blessed Saviour's assurance of pardon and acceptance, to be at peace. I wish to find no rest short of rest in him. Let us both look up to that heaven where our Saviour dwells, and from which he is showing us the attractive face of our blessed and happy child, bidding us prepare to come to her, since she can no more visibly come to us.— *William Wirt.*

> YET cease, my soul! Oh! hush this vain lamenting;
> Earth's anguish will not alter Heaven's decree.
> In that calm world whose peopling is of angels,
> Those I call mine still live and wait for me.
> They can not re-descend where I lament them;
> My earth-bound grief no sorrowing angel shares;
> And in their peaceful and immortal dwelling,
> Nothing of me can enter but my prayers!
> If this be so, then that I may be near them,
> Let me still pray, unmurmuring night and day.
> God lifts us gently to his world of glory,
> Even by the love we feel for things of clay.
> Lest in our wayward hearts we should forget him,
> And forfeit so the mansion of our rest,
> He leads our dear ones forth, and bids us seek them
> In a far distant home among the blest.

So we have guides to heaven's eternal city;
And when our wandering feet would backward stray,
The faces of our DEAD arise in brightness,
And fondly beckon to the holier way.

"NOT LOST, BUT GONE BEFORE."

FOR are we not apt to grieve over the going down of our friends to the grave, as if they were to be forever hidden in its dark chamber, or as if the bright spark of their immortality had been suddenly quenched? They have gone from us; the horizon of death shuts them out of view; their light of love, of hope, of piety, shines no more upon us, and we shall never again behold them in the flesh. But they are no more lost than the sun is lost, when his red disk rolls down behind the western hills. They are no more extinguished than the burning orb of day is quenched when he sinks beneath the waves of the ocean; for as the sun, leaving us in darkness, still lights up other lands, so our departed ones shine in another sphere of existence still — not lost, not extinguished, but made to glow with a brighter light and a more enduring glory. When, therefore, we stand by their coffins, by their graves, or return sad and heavy-laden to their vacant dwellings, we should not mourn for them as those without hope; we should not give vent to grief, as though they were lost to us altogether. They are hidden, but not lost; removed from our sight, but not extinct. They are still alive, only with a more exquisite vitality—unfettered by sin, unencumbered by flesh, undefiled by the world, dwelling as redeemed spirits in the paradise of God.— W. B. Stevens.

I SHINE in the light of God,
   His image stamps my brow;
Through the valley of death my feet have trod,
   I reign in glory now.

I have found the joys of heaven,
   I am one of the angel band;
To my head a crown of glory is given,
   And a harp is in my hand.

O friends of my mortal hours!
   The trusted and the true,
Ye are walking still through the valley of tears,
   But I wait to welcome you.

CHASTISEMENTS.

By some other demonstrations than the dark demonstrations of the storms of sorrow, we know the benevolence of God; and as afflictive dispensations do not *spring from the dust*, but are appointed of God, we have reason to deem them disciplinary—a part of the discipline of his love. His entire benevolence is not incompatible with all the earthly sufferings which so often afflict us to behold, and sometimes almost crush us to bear. How it is that his infinite power should not be wielded by his infinite benevolence to shield us from harm, that he should so often and so deeply embitter our cup, since his benevolence is infinite and pure, must ever remain to us here as one of those deep and dark things of God which no human wisdom can penetrate. As we gaze at the darkness of the cloud that covers us, nothing will answer our purpose but that childlike faith which recognizes it as God's cloud, and thinks of the light which beams in our Father's house beyond it.— *I. S. Spencer.*

AND if it should be, then, Thy will
A cloud should on the future be,
The bow of promise spans it still;
I will believe—I need not see!

Even if the darkness should appear
Too deep for faith as well as sight;
If I am thine, thou wilt be near,
And take me to thy heavenly light.

But, O my Lord! in life's highway
I crave the sunshine of thy face;
And every moment of the day
I need thy strong, supporting grace.

DEATH OF A DAUGHTER.

THIS day two months the Lord delivered my Jessie, HIS JESSIE, from a body of sin and death, finished the good work he had began, perfected what concerned her, trimmed her lamp, and carried her triumphing "through the valley of death." I rejoiced in the Lord's work, and was thankful that the one, the only thing I had asked for her, was now completed. I saw her delivered from much corruption within, from strong and peculiar temptation without. I had seen her often staggering, sometimes falling under the rod; I had heard her earnestly wish for deliverance from sin, and when death approached, she was more than satisfied; said she had been a great sinner, but she had a great Saviour; praised him and thanked him for all his dealings with her — for hedging her in, for chastising her, and even prayed that sin and corruption might be destroyed, if the body should be dissolved to effect it. The Lord fulfilled her desire, and I may add, mine. He lifted upon her the light of his countenance; revived her languid graces; increased her faith and hope; loosed her from earthly concerns, and made her rejoice in the stability of his covenant, and to sing: "All is well, all is well; good is the will of the Lord." I do rejoice, I do rejoice; but, O Lord! thou knowest my frame. She was my pleasant companion, my affectionate child; my soul feels a want. Oh! fill it up with more of thy presence; give yet more communications of thyself.

Let me then gird up the loins of my mind, and set forward to serve my day and generation, to finish my course. The Lord will perfect what concerns me; and when it shall please him, he will unclothe me, break down these prison-walls, and admit me into the happy society of his redeemed and glorified members.—*Isabella Graham.*

> And yet I live to faint and quail
>   Before the human grief I hear;
> To miss thee so, then drown the wail
>   That trembles on my lips in prayer.
> Thou praising, while I vainly thrill;
>   Thou glorying, while I weakly pine
> And thus between thy heart and mine
>   The distance ever widening still.

Two months of tears to me—to thee
   The end of thy probation's strife,
The archway to eternity,
   The portal of immortal life.
To me the pall, the bier, the sod;
   To thee the palm of victory given.
Enough, my heart; thank God! thank God!
   For thou hast reached thy home in heaven.

THE AGED, LOOKING BACK TO YOUTH.

Not as the leaves of autumn, all at once, have the generations of man fallen and disappeared from my sight; but one by one they steal away, and others fill their places, till the last survivor, like myself, withering amidst his fresh and vigorous successors, falls alone, as I shall do, unlamented and almost unobserved. Could a vision be seen of the many who formerly loved me, of all with whom I was once intimately associated, how numberless would they appear! But now, like a vast field of battle strewed over with the dead, the world lies desolate around me. In a home once peopled with parents, sisters, brothers, and friends, I hear only the echo of my own solitary footsteps; no outstretched hand or smiling countenance welcomes my return, no familiar voice greets my ear—my generation has passed away.—*Catherine Sinclair.*

When at eve I sit alone,
Musing on the past and gone,
While the clock, with drowsy finger,
Marks how long the minutes linger,

And the embers, dimly burning,
Tell of life to dust returning—
Then my lonely chair around,
With a quiet, mournful sound,
With a murmur soft and low,
Come the ghosts of long ago.

One by one I count them o'er,
Voices that are heard no more,
Tears that loving cheeks have wet,
Words whose music lingers yet,
Holy faces, pale and fair,
Shadowy locks of waving hair,
Gentle sighs and whispers dear,
Songs remembered many a year.

THE INTRUSTED JEWELS.

During the absence of the Rabbi Meir, his two sons died—both of them of uncommon beauty, and enlightened in the divine law. His wife bore them to her chamber, and laid them upon her bed. When Rabbi Meir returned, his first inquiry was for his sons. His wife reached to him a goblet; he praised the Lord at the going out of the Sabbath, drank, and asked again: "Where are my sons?" "They are not far off," she said, placing food before him, that he might eat. He was in a genial mood, and when he had said grace, after meat, she thus addressed him: "Rabbi, with thy permission, I would fain propose to thee one question." "Ask it, then, my love," replied he. "A few days ago a person intrusted some jewels to my custody, and now

he demands them. Should I give them back to him?"
"This is a question," said the Rabbi, "which, my wife,
I should not have thought it necessary to ask. What!
wouldst thou hesitate or be reluctant to restore to every
one his own?" "No," she replied, "but yet I thought
it best not to restore them without acquainting thee
therewith." She then led him to the chamber, and step-
ping to the bed, took the white covering from the dead
bodies. "Ah! my sons, my sons!" loudly lamented
their father; "my sons! the light of my eyes and the
light of my understanding. I was your father, but you
were my teachers in the law." The mother turned
away and wept bitterly. At length she took her husband by the hand, and said: "Rabbi, didst thou not
teach me that we must not be reluctant to restore that
which was intrusted to our keeping?"

WHAT bliss is born of sorrow!
  'Tis never sent in vain;
The heavenly Surgeon maims to save,
  He gives no useless pain.

Our God, to call us homeward,
  His only Son sent down;
And now, still more to tempt us there,
  Has taken up our own.

THE DEPARTED.

THOUGH better informed as to the objects of our love
than they who lingered about the deserted tomb of the
Saviour, and were asked, "Why seek ye the living

among the dead?" we nevertheless find ourselves, in our thoughts, searching for them, so difficult is it at once to feel that they are wholly and forever departed. There is an affecting and beautifully simple illustration of our thoughts and feelings, in this respect, in the search which was made for Elijah, after his translation. Fifty men of the sons of the prophets went and stood to view afar off, when Elijah and Elisha stood by the Jordan. Elisha returned alone, and those men could not feel reconciled to the loss of their great master. They were not persuaded that he had gone to heaven, no more to return. They sought leave to seek and to recover him. "Peradventure," they said, "the Spirit of the Lord hath taken him up, and cast him upon some mountain, or into some valley." Elisha peremptorily refused to grant them leave. They were importunate ; and when at last it would, perhaps, seem like obstinacy in him, or like jealousy of their superior love for Elijah, to forbid the search, which, at the worst, would only be fruitless, he yielded. Three days they explored the valleys, ransacked the thickets, groped in the caves, traversed hills, followed imaginary trails and footprints, but found him not. When they came again to Elisha, he said unto them : " Did I not say unto you, Go not ?"

Suppose that those "fifty strong men" had found Elijah, or in any way could have prevented his translation to heaven. With exultation they would have led him back across the Jordan, to the company of his friends, amidst the thanksgivings of the people. But, alas ! for the prophet himself, this would have been his loss, even had it proved to be their gain. The opening

Jordan, cleft in twain by his rapt spirit, pressing its
way to the skies, had returned to its course; and now
the fords of the river, with its rocky bed, would have
required his laboring feet to grope their way back to
his toil, or the arms of men, instead of the chariots of
fire and horses of fire, would have borne him again to
the dull realities of life. Blind and weak do these "fifty
strong men" seem to us, in searching for this ascended
prophet, this traveler over the King's road in royal state,
one of the only two who might not taste of death.

And while they grow weary and discouraged, the
glorified Elijah was with Abraham, Isaac, and Jacob,
and with Moses, Samuel, and David. To-day our loved
ones in heaven look upon him, and say, as Peter did at
this prophet's visit on Tabor: "Master, it is good for
us to be here." But we, like the "fifty strong men,"
would find them and bring them back; and, like Peter,
would build tabernacles to retain them. The family
circle is gathered together at some birthday or festival,
and, perhaps, we long for the departed, and think that
they long for us; and we would bring them back, and
place them in their deserted chairs. We are "strong
men" in the power of grief, and in our wishes; and the
search for Elijah is the counterpart of our vain desires
and most unreasonable sorrow.—*Nehemiah Adams.*

> We miss thee from the band so dear,
>   That gathers round our hearth;
> We listen still thy voice to hear,
>   Amid our household mirth.
> We gaze upon thy vacant chair,
>   Thy form we seem to see;

We start to find thou art not there,
  Yet joy that thou art free.

A thousand old familiar things,
  Within thy childhood's home,
Speak of the cherished, absent one,
  Who never more shall come.
They wake, with mingled bliss and pain,
  Fond memories of thee;
But would we call thee back again?
  We joy that thou art free!

THE "ELECTRIC CHORD" OF ASSOCIATION.

BELGIUM! I repeat the word now, as I sit alone at midnight. It stirs my world of the past like a summons to resurrection: the graves unclose, the dead are raised; thoughts, feelings, memories that slept, are seen by me ascending from the clods, haloed the most of them; but while I gaze on their vapory forms, and strive to ascertain definitely their outline, the sound which wakened them dies, and they sink, each and all, like a light wreath of mist, absorbed in the mold, recalled to urns, resealed in monuments!—*Charlotte Brontë.*

FAST as its breathings rose, like blissful clouds,
  Fair phantoms upward on the vapor curled;
Sweet resurrections, breaking from their shrouds,
  Stood pale before me, like an ancient world.

To me the veil of time was rent in twain—
  Eve changed to morn, the morn into the sun;
Behind the cloud of days I saw again
  A feast, a bridal, and the first of June.

The very music seems to hover by
  The songs we sang together in the bower;
I hear that ghostly music with a sigh—
  The lips are dust that rained the silver shower.

INFANTS IN HEAVEN.

IF God sees proper in mercy to relieve any of our race from the toils and responsibilities of earth, by taking them to heaven in infancy, we should glory in his grace. They leave their loved ones without the pangs of parting. They yield to the embraces of death without knowing that it is a penalty. They lie down in the grave without any thoughts of its loneliness. They enter the eternal world without any dread of its retributions. They fly back to the bosom of their Father with the same innocent confidence as they were wont to fall into the arms of their earthly parents. So are they forever with the Lord! They have obtained rest without weariness; they have been victorious without a conflict; they are saved without a probation.—*Harbaugh*.

GOD took thee in his mercy,
  A lamb untasked, untried;
He fought for thee,
He gained the victory,
  And thou art glorified.

GIRLHOOD AND OLD AGE.

Who would believe that the faded, worn-out being I now am could ever have been, or even claimed kindred with the sanguine, joyous, happy girl once surrounded within these very walls by parents, friends, companions, and even by lovers — all, all now crowded into their silent graves! How many faces, remembered by none but myself, are yet present to me, vivid as they were in by-gone times, with life and gayety! I have lived to be the last depository of their memories, the last on this earth who remembered their countenances, who had shared in their thoughts, or would drop a tear over their graves. Yes, of all who rejoiced with me in joy or mourned with me in sorrow, I alone remain. Oh! how I sometimes long to behold but one living being who could remember the days that I remember!— *Catherine Sinclair.*

It was not thus when dreams of love and laurels
    Gave sunshine to the winters of our youth,
Before its hopes had fallen in fortune's quarrels,
    Or time had bowed them with its heavy truth;
Ere yet the twilight found us sad and lonely,
    With shadows coming when the fire burns low,
To tell of distant graves and losses only,
    The past that can not change and will not go!

I HAVE BEEN LIKE ONE IN A FEVER.

I HAVE been like one in a fever, attended at times with a strong delirium. I begged hard that I might be spared, but He meant a cure and pierced my heart. Oh! how slender, how brittle the thread on which hang all my earthly joys!

When I find my joys packed up and gone, my heart slain, the delight of my eyes taken away; when I recollect who has gone before her, who is following, and what remains for the world to offer, my heart cries, " I loathe it; I would not live alway," I thank God that I also am to go. I did not know how much my heart was bound up in the life of a creature; when *she went, nothing seemed left.* I have often prayed: " Lord, soften my heart, humble my pride, destroy my levity." I knew enough of his way to fear the means, and he has in mercy toward me regarded my soul more than my feelings; and now I can say: " Lord, to whom shall I go but to thee?"—*Richard Cecil.*

No flowers, no garlands gay?   All blasted?
     All wasted?
But as I raved and grew more fierce and wild,
     At every word,
Methought I heard one calling: " Child!"
And I replied: " My Lord!"

DEATH OF A FATHER.

I AM at home again. I have been home a long time. There is a long interval since my last entry and the present, and a longer period in my life. I have endured the greatest affliction that ever could befall me in that space of time. When last I wrote in this brief record of my daily employments, I was happy; I had no cares but those I made for myself, no reasonable wishes ungratified, and I was sheltered from every evil thing in the sweet, strong refuge of my father's love. Now how changed! It is the same house, the same room, nothing around me is altered; but in one fearful day all earth's hopes, peace, enjoyment, protection have left me forever. I AM FATHERLESS! When the decree went forth that he should be translated, if it had been done gently and by degrees, instead of suddenly, roughly wrenching away without a word of warning all that made life desirable, we might have borne it better. But such was not God's will. In the morning the tall tree stood without one token of decay, and bore up its feeble companions with a strong support, and at night the poor ones lay crushed and bleeding — their prop had been cut down.

The trials of past years, and they were neither few nor slight, are all swallowed up in this. We bore them patiently, cheerfully, because we had hope. Now we have none. The grave can not give up its trust; the precious clay will not revive at our bidding, the dear voice answers not our passionate invocations—we are alone.—*Miss Griggs.*

> I CALLED—to call what answers not our cries—
>   To stand by that we love, unseen, unheard ;
> In the deep passion of our tears and sighs,
>   To see but some cold, glittering ringlet stirred !
> And in the quenched eye's fixedness to gaze,
> Searching all vainly for the soul's bright rays :
> This is what waits us ! *Dead !* with that chill word
> To link our bosom names ! For this we pour
> Our souls upon the dust, nor tremble to adore !

DEFECTION IN FRIENDS.

BUT a trying time came—a bleak, cold north wind and a very sharp, piercing frost ; like leaves in autumn, down fell the promising bloom. Thou art mourning for the loss of living friends. They have forsaken thee. Old connections, as dear to thee as thine own soul, are broken. Persons whom thou hast known from thy childhood, and with whom thou hast grown up in strict friendship, are now thine enemies, and become so without any offense or fault of thine. 'Tis even so. It was not an enemy that reproached me—then I could have borne it ; neither was it he that hated me, that did magnify himself against me—then I would have hid myself from him ; but it was thou, mine equal, my guide, and mine acquaintance — mine own familiar friend in whom I trusted, which did eat of my bread.

As other ties are dissolved, thy heart will knit closer to thy divine Friend.—*Romaine.*

> Is it not now the north wind finds us shaken
> By tempests fiercer than its bitter blast,
> Which fair beliefs and friendships too have taken
> Away like summer foliage as they passed,
> And made life leafless in its pleasant valleys,
> Waning the light of promise from our day ;
> Fell mists meet even in the inward palace,
> A dimness not like theirs to pass away.

THE DREAM.

WEEKS passed on after her death, and although I did not "refuse to be comforted," yet I seemed to be beyond the reach of consolation. I would sit for hours thinking that God had dealt severely with me, wondering why he had given her to me for so SHORT a time, wondering why he had made her so lovely and attractive, just to make me dote on her so fondly, and wondering why he had sent her AT ALL when I was so very happy before she was given. And then the dreary thoughts I would have about the little body in ruins instead of thinking of the spirit in glory! I would sit and murmur to myself: "O that sweet, joyous creature—shut up in the dark vault, where no ray of light ever comes! O the little sleeper! not in the comfortable crib, but under the coffin-lid, with the little waxen hands so cold and still that used to be so busy!"

How I would sit and watch the snow falling, and feel agonized by the thought that she must be suffering with the cold; and then when the high March winds would rave around the house at night, I could not get rid of the feeling of distress that she would be awakened and feel alarmed at finding herself all alone, forgetting that she had fallen into that "sleep" which nothing could break but the Archangel's trump.

Three long dreary weeks passed by me under this cloud, and I all that time was murmuring at what I thought hard dealings; but as a tender parent listens sorrowfully and patiently to the wild ravings of his sick child in delirium, even so God stood by me in sympathy, and bore with me in love till the fever of sin and discontent had passed away. One night I had a dream, and oh! how differently I regarded the removal of my child, just as the natural landscape, when seen through a violet-colored glass, looks dull and gray and wintry; while that same landscape, when viewed through another shade, seems bright and glowing and gorgeous; and so the dream, through God's grace, had given another coloring to "God's ways" and my child's happiness, and my thoughts in sleep had brought me to realize the blessed truth, "that he doth all things well."

I dreamt I was standing by a low log cabin, with a dreary lake or cypress swamp spread out before me. I held Lilian in my arms. I was very unhappy, feeling there was a strong necessity on me to carry the child over the lake, but I was afraid to venture. Strong presentiments of evil weighed me down, and I lingered till sunset, watching the long shadows on the grass and the

cypress-trees as they stretched their low branches over
the gloomy lake. As the sun sank out of sight, great
fear came over me, and I hastened down a path which
led to the swamp. A corduroy road, covered with
moss, stretched across the lake. As I placed my foot
on the corduroy road, slippery with moss, the road
sank down into the water, and hundreds of snakes,
from every root and branch of the cypress-trees, raised
their heads and hissed and reached toward us. I rush-
ed back up the low bank, terrified and trembling, hardly
able to hold the clinging child in my exhausted arms.
While standing there, still feeling impelled to carry Lilian across the swamp, I saw a young man at a distance,
with his back toward me, and thought it was my brother. As I approached him, I called out several times
in great distress : "Oh! help me to take Lilian over
the swamp; oh! help me to carry the child over."
Just as I reached the spot where he stood, he turned
at the sound of my voice — it was Jesus Christ! He
held his hands out lovingly to my child, and I placed
her in the arms of the divine Saviour who had said :
"Suffer little children to come unto me." He passed
me, walked toward the lake, and I followed them.
The night was coming rapidly on; the swamp looked
gloomier than ever; the snakes still hissed and reached
toward us from all their coiling places. The way
seemed very long as I toiled over the slimy moss; but
the little arms were clasped contentedly around the
Saviour's neck, and the dear, dear face looked down on
me over his shoulder, and I held on to our Guide.

When we reached the opposite shore and I knew that she was safe, my joy was so intense that I awoke.

And may I not think that God sent me that dream—sent it to the poor ignorant sinner to convince her that "infinite wisdom never makes a mistake," that the heavenly Father always chooses what is BEST for his short-sighted, erring child? And the dream has had its mission; for me, I have never since felt that any of his dealings were HARD. I have never since questioned his loving discipline, though I have been led many times by his providences to pass through fiery furnaces and strong water-floods. And as it regards my feelings and thoughts about Lilian, I am more than satisfied. I never think of her as the little sleeper occupying the dark, gloomy vault, but as the sweet child in the arms of her Saviour, taken from life without being tried by its sorrows, or wearied by its tasks, and taken from sin without struggling against its temptations or soiled by its defilements.—*A. N.*

THAT voice of music filled my ears,
I thought her mine through long, long years,
At day-dawn missed her, blind with tears:

But now those faithless tears are dried;
Here at my calling could she glide,
I would not call her to my side.

From visions of her Saviour—King,
From blisses past imagining,
Dare love like mine its dear one bring

Where sin would soil my snow-wreath fair—
That dear voice moan in earth's despair?
Oh! no, I would we *all* were there!

THE AGED ON THE BANKS OF THE RIVER.

AGED believer, you are now standing on the banks of
the river. Fear not, only believe. Remember that one
of the reasons why Jesus Christ manifested himself in
human nature was, for the express purpose of dispelling
that gloom which naturally overspreads the mind as we
look upon the dark waters of death. "Forasmuch as
the children are partakers of flesh and blood, he also
himself likewise took part of the same; that through
death he might destroy him that had the power of
death, that is, the devil, and deliver them who, through
fear of death, were all their lifetime subject to bondage."
Can you say with gladness: "The time of my departure
is at hand: I have fought a good fight, I have finished
my course, I have kept the faith; henceforth there is
laid up for me a crown of righteousness, which the Lord,
the righteous Judge, shall give me at that day"? Thank
your Saviour for this glorious hope, this hope which is
as an anchor of the soul, sure and steadfast, for he is its
author and its bestower. It is because he has abolished
death, and brought life and immortality to light through
the Gospel, that you are now enabled to look forward
with composure to your conflict with your last foe.
Well may you rejoice, for your life is hid with Christ in

God, and you are safe forever. Safe amidst the infirmities and perils of old age; safe in the swellings of Jordan; safe when you stand before the solemn judgment-
seat; yes, safe throughout eternity. Nothing in earth
or hell can separate you from the love of God which is
in Christ Jesus, or pluck you from the grasp of your
everlasting Saviour. He upholds and comforts you
now in the evening of life, and by and by, leaning upon
his arm, you shall come down to the river. Not a
ripple shall be on its bosom; its clear waters, shining
in heaven's own light, shall allure to the crossing. His
feet shall but touch the stream, and lo! a way for the
ransomed to pass over.—*Life's Evening.*

Thy life-cruise is ending,
  White crest of each wave,
With swifter rush tending,
  Home's ramparts to lave:
Then fear not the blending
Of cloud, reef, and foam—
Heart well-nigh home.

Heart, therefore, lay all
  Low at His feet;
Years of betrayal,
  Service how fleet!
Waiting there thine arrayal,
Meet for heaven's dome—
Heart, well-nigh home!

WHAT IS DEATH TO THE BELIEVER?

WHAT is death to the believer? It is the beginning of eternal life. It is the coronation-day of one who will reign with Christ forever. It is only opening the door to let a prisoner of hope out into the pure air and sunlight of heaven. It is sending a weary pilgrim home to his everlasting rest.

> SEEMS cry of the night-owl dreary?
> Dawn cometh to lift the cloud,
> Then for watchers no longer weary
> Will song of the lark be loud.
> Of the lark! To the soul far sweeter
> Than ever morn-music rose,
> Shall the welcome of Jesus greet her,
> Escaping from sin's last woes.

NOW LOOK HIGHER.

OH! the anguish we sometimes get from the things that once delighted us! And oh! the blessedness from that anguish, too! As long as we can get sweetness and unalloyed sweetness from any earthly object, we shall never turn from it; such things are too rare in the earth, and we too hungry. God, therefore, after a little, lays gall and wormwood on the thing we love, and more and more of it, till its sweetness goes, and at last we are afraid of it. But we want it still, for it is

still sweet to us; but he says, "No, you shall have it
no longer;" and then comes a worm and withers our
gourd, friends are alienated, breaches are made in our
families, graves are opened, and houses and hearts left
desolate. We would not tear our soul from that object;
God therefore tears that object from us, and says when
he has done it: "Now look higher."—*Charles Bradley.*

Go and tell Jesus, when thine eye hath seen
　　Dear hopes destroyed by the tyrant Death ;
When reeds thou lovest pierce the hands that lean—
　　Hear what he saith.

Go and tell Jesus.　In his wisdom lie
　　All stores of solace.　When rude gales increase,
Ask, and his love shall pour on passions high
　　The oil of peace.

THE CHILD IS DEAD.

It is hard to believe it, that we shall no more hear
the glad voice or meet the merry laugh that burst so
often from its glad heart.

It was a pleasant child, and to the partial parent
there are traits of loveliness that no other eye may see.
It was a wise ordering of Providence that we should
love our own children as no one else loves them, and as
we love the children of none besides. And ours was a
lovely child. You may put away its playthings; put
them where they will be safe. I would not like to have
them broken or lost. Do not lend them to other child

ren when they come to see us. It would pain me to
see them in other hands, as much as I love to see children happy with their toys.

Lay his clothes aside. I shall often look them over.
They will remind me of him as he looked when he was
here.

I shall weep often when I think of him. The little
hand is still and cold, the little heart is not beating
now. To think of the little one laid in its coffin! He
never was in so cold and hard a bed; but he will not
feel it. I hope he was carried to the grave gently! It
is a hard road to the grave. Every jar seems to disturb the infant sleeper; and then to stand by the open
grave! How damp and cold and dark it is! But the
dead do not feel it. There is no pain, no fear, no weeping there. How every clod seems to fall on the heart
as they fill up the grave! Every smothered sound from
it seems to say: Gone, gone, gone! But our child is not
there; his dust, his precious dust is there, but our
child is in heaven; but I can not but think of the form
that is here moldering among the dead. It will be a
mournful comfort to come at times to his grave and
think of the child that was once the light of our house
and the idol of my heart. And it is beyond all language to express the joy in the midst of tears, to feel
that my sin, in making an idol of my child, has not made
that infant less dear to Jesus. Nay, there is even something that tells me the Saviour called the darling from

me, that I might love him more. Dear Saviour, as thou hast my lamb, give ME, too, a place in thy bosom.

"Ir's only a l.ttle grave," they said,
"Only just a child that's dead;"
And so they carelessly turned away
From the mound the spade had made that day.
Ah! they did not know how deep a shade
That little grave in our home had made.

I know the coffin was narrow and small—
One yard would have served for an ample pall;
And one man in his arms could have borne away
The rosewood and its freight of clay;
But I know that darling hopes were hid,
Beneath that little coffin-lid.

'Tis a little grave; but oh! have care,
For our precious child was buried there;
And ye, perhaps, in coming years,
May see, like her, through blinding tears,
How much of light, how much of joy,
Is buried up with an only boy!

HEAVEN HAS ATTRACTIONS.

WHERE is he who used to lisp, "father—mother," thy child? Passing out of your hands, passed he not into those of Jesus? Yes, you suffered him. If any other than Jesus had said, "Suffer them to come to me," you would have said, no. Death does not quench those recently struck sparks of intelligence. Jesus is not going to lose one of those little brilliants. All shall be in his crown. Perhaps thou hast a brother or a sister there; that should draw you to heaven. Perhaps a *mother*—she whose eye wept while it watched over thee, till at length it grew dim, and closed. Perhaps one nearer, dearer than child, than brother, than sister, than mother, the nearest, dearest is there. Shall I say who? Christian female, thy husband. Christian father, the young mother of thy babes. He is not, she is not, for God took them.— *William Nevins.*

> BELOVED! where hast thou been these years?
> What hast thou seen?
> What visions fair, what glorious life,
> Where thou hast been?
>
> The vail! the vail! so thin, so strong,
> 'Twixt us and thee—
> The mystic vail! when shall it fall,
> That we may see?

THE FEAR OF EVIL.

"I SHALL not want." Simple as these words are, how few of us could feelingly utter them! They indicate a state of mind for which our hearts often and greatly long, but which we find hard to attain, and when attained, harder still to keep, a being careful for nothing, a state of quietness and repose. The man who wrote it seems to have been without an anxiety or a fear. "I shall not want," he says at first, and then a little after: "I will fear no evil." "The Lord is MY

Shepherd; I can look up to him as MINE." And this connecting of a gracious God with ourselves is necessary for us before we can have any abiding peace in him. A believing view of God, as in Christ Jesus, a gracious God, will, I know, save my guilty soul when I die; but it will not of itself quiet my troubled spirit while I live. I must see his favor and mercy reaching to me, his peculiar mercy, the favor he bears to his chosen. I must feel myself to be an object of it, embraced by it, under its influence and operation, and then I can rest, then I can say, "Abba, Father;" then I know I am safe. Place me then in the wildest desert on the globe, amidst perils out of number, in desolation and darkness, do with me what you will, I can say, and say it with as much confidence, blessed be God, as though I were in heaven: "I will fear no evil; I shall not want." How can I? There is the omnipotent God, my Shepherd, to protect me, and there is the same God, with all his riches in glory, my Shepherd, to feed me.—*Charles Bradley*.

THE fear of evil! 'Tis an evil thing,
  For in thy presence, that all-shadowing tree,
The heart should build her nest, and bird-like, sing,
  Leaving the morrow's care a charge for Thee;
Not quail, as lonely hare
Sinks down in sombre lair,
  Hearing far bugles, though the woods are free.

THE MISSIONARY'S PARENTS.

THE intelligence contained in your letter was not un-
expected. Our father had attained to a great age, lack-
ing only five days of being eighty-six years old. He
was full of days, but more full of faith and of the Holy
Ghost. Though I can look back some forty-five years
or more, I can not look back to the year when he was
not living a life of faith and prayer and self-denial, of
deadness to the world, and of close walk with God.
Though his means of grace were limited, yet meditating
day and night on God's law, his roots struck deep, and
he was like a tree planted by the rivers of water, whose
leaf is always green, and whose fruit is always abundant.
Whoever saw him riding on horseback would, if he
kept himself concealed, be almost sure to see him engaged in prayer. Whoever would work with him in
seed-time or harvest would find his thoughts as actively
employed above as his hands were below. Whoever
of the Lord's people met him, by day or by night, at
home or abroad, alone or in company, would find him
ready to sit down with them in heavenly places, in order
to comprehend " what is the length, and breadth, and
depth, and height " of the love of Christ. Being the
youngest of the family, you can have but an indistinct
recollection of two small rooms and a garret, floored
with loose and rough boards, where twelve of us were
born, and of the small clump of apple-trees before the
door, where your elder brothers and sisters played in
the days of their thoughtless childhood. There, with

no lock to any door, and no key to any trunk or drawer
or cupboard—there, where, as I am told, nothing now
remains but an old cellar, which may even itself, long
before this, have been filled up—there our godly father
prayed for us with all prayer and supplication in the
Spirit; there, on every Sabbath eve he asked us those
solemn, important, and all-comprehensive questions
from the catechism; and there, with eyes and heart
raised to heaven, he used to sing, to the tune of old
Rochester:

> God, my supporter and my hope,
>   My help, forever near;
> Thine arm of mercy hold me up,
>   When sinking in despair.

And there, too, our mother, of precious memory,
though, as she died when you were but six months old,
you remember her not, there she lived a life of poverty,
patience, meekness, and faith. There she used to sit
and card her wool by the light of the pine-knot, and
sing to us those sweet words:

> Hovering among the leaves, there stands
>   The sweet celestial Dove;
> And Jesus on the branches hangs
>   The banner of his love.

And there, too, almost thirty-four years ago, we assembled early one morning in her little bedroom to see
her die. Her peace was like a river; she was full of
triumph, and she was able to address to us words of
heavenly consolation, till she had actually crossed over

into shallow water, within one minute of the opposite
banks of the Jordan, *heaven and all its glories full in
view.*

But before I close I must say something more of the
early habits and character of our venerable father. The
little farm he once possessed, if it were not all *ploughed*
over, was, I am confident, almost every foot of it prayed
over. He served three years in the Revolutionary War,
and I was struck with the fact you communicated of its
being early on the morning of the memorable fourth of
July, amidst the roaring of cannon, that he slept in
peace. And though to his children he left no inheritance—no, not so much as one cent—yet, in his godly
example and prayers, he has left them the very richest
legacy which any father ever bequeathed his children.

It is a rare privilege we have all enjoyed in being
descended from such parents. They were the children
of the great King. They belonged to the royal family.
They daily walked abroad with the conscious dignity
of heirs to a great estate, even an incorruptible inheritance; and now they have gone to sit down with Christ
on his throne.— *William Goodell.*

My boast is not that I deduce my birth
From loins enthroned, and rulers of the earth;
But higher far my proud pretensions rise—
The son of parents passed into the skies.

THE GLORIFIED BODY.

THE glorified body! how immeasurably will it transcend in physical and moral beauty the old earthly tabernacle! "Sown in corruption, raised in incorruption; sown in weakness, raised in power; sown a natural body, raised a spiritual body." Glorious body indeed! without sin, without pain, without weakness, or weariness, or infirmity. The grave will not be permitted to efface the memorials of the past, and destroy our personal identity. The resurrection body will wear its old smiles of love and tenderness. The features of my buried friend I shall recognize again. The beaming face of cherished affection shall bear the old impress of earth. No change but this, that the shifting tent is transmuted into a "building of God," reared of permanent and imperishable materials, a bodily structure that shall know no decrepitude—smiles that shall never die.—*Grapes of Eschol.*

> BUT if the Spirit's blessedness be such,
> What of the body? Mortal tenement,
> (Mortal and frail,) yet loved, oh! yes, how loved!
> Each feature penciled as with living light
> On the soul's tablets, ineffaceable,
> Smiles that can never die! Say, can it be
> That all now left of these is memory?

LOSS OF A HUSBAND.

You that knew us both, and how we lived, must allow I have just cause to bewail my loss. I know that it is common with others to lose a friend; but to have lived with such a one! it may be questioned how few can glory in the like happiness, so consequently lament the like loss! My heart mourns, too sadly I fear, and can not be comforted, because I have not the dear companion and sharer of all my joys and sorrows. Can I regret his quitting a lesser good for a bigger? Oh! if I did steadfastly believe, I could not be dejected, for I will not injure myself to say, I offer my mind any inferior consolation to supply this loss. I strive to reflect how large my portion of good things has been, and though they have passed away, no more to return, yet I have a pleasant work to do, dress up my soul for my desired change, and fit it for the converse of angels and the spirits of just men made perfect; amongst whom my loved lord is one; and my often-repeated prayer to my God is, that if I have a reasonable ground for that hope, it may give a refreshment to my poor soul.

The future part of my life will not, I expect, pass as perhaps I would just choose. Sense has been long enough gratified; indeed, so long, I know not how to live by faith; yet the pleasant stream that fed it near fourteen years together being gone, I have no sort of refreshment but when I can repair to the fountain of living waters.

I am entertaining some thoughts of going to that
now desolate place, Straton, for a few days, where I
must expect new, amazing reflections at first, it being
a place where I have lived in sweet and full content;
considered the condition of others, and thought none
deserved my envy. But I must pass no more such days
on earth; I can not recover what was a perpetual bliss
to me here. A flood of tears is ever ready when I per-
mit the least thought of my calamity.

'Twas, Doctor, an inestimable treasure I did lose,
and with whom I had lived in the highest pitch of this
world's felicity. I was too rich in possessions whilst I
possessed him; all relish now is gone. I bless God for
it, and pray more and more to turn the stream of my
affections upward. The new scenes of each day make
me often conclude myself very void of reason, that I
still shed tears of sorrow, and not of joy, that so good
a man is landed safe on the happy shore of a blessed
eternity. Doubtless he is at rest, but I find none without him, so true a partner he was in all my joys and
griefs.—*Lady Rachel Russell.*

'Tis ever thus, 'tis ever thus, that when the poor heart clings
With all its finest tendrils, with all its flexile rings,
That goodly thing it cleaveth to, so fondly and so fast,
Is struck to earth by lightning, or shattered by the blast.

'Tis ever thus, 'tis ever thus, when hope hath built a bower
Like that of Eden's, wreathed about with every thornless flower,
To dwell therein securely, the self-deceiver's trust,
A whirlwind from the desert comes, and "all is in the dust."

REST IN DEATH.

DURING the last hour of your sainted brother's life, Mr. Ranney bent over him, and held his hand, while poor Panassah stood at a little distance weeping bitterly. The officers did not know what was passing in the cabin, till summoned to dinner. Then they gathered about the door, and watched the closing scene with solemn reverence. Now — thanks to a merciful God!—his pains had left him; not a momentary spasm disturbed his placid face, nor did the contraction of a muscle denote the least degree of suffering; the agony of death was passed, and his wearied spirit was turning to its rest in the bosom of the Saviour. From time to time he pressed the hand in which his own was resting, his clasp losing in force at each successive pressure; while his shortened breath — though there was no struggle, no gasping, as if it came and went with difficulty—gradually grew softer and fainter, till it died upon the air, and he was gone. Mr. Ranney closed the eyes, and composed the passive limbs.— *Emily Judson.*

Two hands upon the breast,
    And labor's done;
Two pale feet crossed in rest,
    The race is won;
Two eyes with coin-weights shut,
    And all tears cease;
Two lips where grief is mute,
    Anger at peace.

GONE HOME.

THEY traveled by the express-train, and got so quickly over the ground, that soon they were within a few miles of their journey's end, and were beginning to talk of those they would see there. Their father asked them to make choice of a psalm to repeat to him. The elder repeated the First Psalm, his little brother the One hundred and twenty-first, each choosing his favorite, and then, in concert, the Twenty-third. They had not very long finished the last verse—

> "Goodness and mercy all my life
> Shall surely follow me;
> And in God's house for evermore
> My dwelling-place shall be;"

when the train, which was going very fast, began to shake from side to side, in a way that alarmed their parents; but it did not frighten the boys much — perhaps they thought they would be the sooner home. And so they were. Suddenly the engine went off the rails; there was a tremendous crash, and in a moment the youngest brother WAS HOME! — the happy spirit was in the "Father's house;" it was only the body of clay that was lying on the bank of the railway. The eldest lingered patiently for thirty-six hours, as if uncertain whether to remain with his beloved parents, or to join his little brother; but he, too, WENT HOME, which was far better, for

> "In God's house for evermore
> Their dwelling-place shall be."

—The Way Home.

A SPRING-DAY journey, such it seemed, to end when night should
come :
A few more miles, another hour, and they should reach their home ;
So nearer, near, when suddenly the angel swerved his hand
Aside from every earthly goal, due for the eternal land.

He swerved aside, because he saw heaven's gateway arching blue ;
One moment's breath, and joyfully the children are let through,
Their spring-day journey at an end, its perils and alarms,
For Jesus on the threshold stood, and clasped them in his arms.

Bear up, brave mother, strong in faith ; bear, father, stricken sore ;
Your little ones are housed and home — what could you wish them
more ?
The voices that are silent here, are singing gladly there,
Or asking God to comfort you, in some sweet, childish prayer.

DEATH OF A MOTHER.

My own dear mother has died ; and when I utter
that expression, and remember what she was to me
from my childhood till her last breath of life, no words,
I am sure, could paint the traces of emotion that come
over me. I expected her death ; I knew it must be
near ; and yet anticipation has not made it a reality
for which I was prepared. Fond of her family, devoted to them, self-sacrificing and ever-faithful, she spared
no pains, shrank from no labor, and shunned no care or
hardship which was demanded for the good of her
family. Though timid by nature, and more inclined
to despondency than hope, she met the cares of a nu-
merous family and the troubles of a changeful life

without complaint or repining. She took the trials of her children as her own trials, adopted their sorrows as her own, and whenever she could, she shielded them from harm by the ready exposure of herself. She was governed by her Bible, conscientious in every thing. Her body now rests on the banks of the Cattaraugus, and the tie which bound her children together and made them feel as one family, is severed forever. Though I anticipated her death and knew it could not be far off, yet I did by no means expect it to impress me as I find it does. I seem now to be cut loose from all that went before me; I seem to have done with all the past, and to be compelled to turn all my thoughts to the future — from my parents to my children — from the generation that went before me to the generation that shall come after me. As long as my mother lived I could be a child. Though I could not think of her any longer as one to lean upon, I *could* think of her as one to love, and think of her, too, as one to lean upon me. I endured and bore up on her account at times when nothing but the thought of her kept me from despair. ONE, at least, would honor me, do me justice, prize me; to ONE, at least, I might be useful.—*I. S. Spencer.*

> My mother's voice! how often creeps
> Its cadence on my lonely hours,
> Like healing sent on wings of sleep,
> Or dew to the unconscious flowers.
> I can forget her melting prayer,
> While leaping pulses madly fly,

But in the still, unbroken air
Her gentle tone comes stealing by,
And years and sin and manhood flee,
And leave me at my mother's knee!

DEATH OF OUR INFANTS.

How beautiful they were! beautiful, even beneath the coffin-lid, with hands folded peacefully, with brow like molded wax, with eyes closed as in sleep.

We miss them every where! We see them every where! Does not every object in the house and around us bring to us thoughts of them? We seem to see them again, when a hasty search-errand to the drawer exposes to our view the clothes and playthings which they left behind. We close it, and weep as we go away.—*Harbaugh.*

I know that a mother stood that day
With folded hands by that form of clay;
I know that burning tears were hid
'Neath the drooping lash and the swollen lid;
And I know her lips and cheek and brow
Were almost as white as baby's now.
I know that some things were hid away—
The snow-white frock and the wrappings gay,
The little sock and the half-worn shoe,
The cap with its plume and tassels blue,
And an empty crib, with its covers spread,
As white as the face of the guileless dead.

DYING GRACE.

You fear to cross its deep, deep waters; you shrink from the strange, and, it may be, the stormy passage to eternity. You say: Oh! if I could but reach the celestial city without having to cross the stream of death! God knows your frame; he remembers that you are dust, and feels the tenderest parental compassion for those who fear him; and therefore you may be assured that the trials which his love ordains, whether in life or in death, are NECESSARY trials, and he will give you support under *them*. His grace is sufficient for you as well as for others. Oh! trust yourself to him; repose with confidence upon his promises; and believe that in a dying hour your succor shall be equal to your need. Do not test your preparedness for that hour by the strength and comfort which you now possess, but by the solemn engagement which Christ has made never to leave nor forsake you. He is with you now, to help you glorify him by your life; when death comes, he will be with you then, and help you glorify him by your death. Dying grace will not be vouchsafed until a dying hour. You do not want it now, but it will be abundantly vouchsafed then. Wait for it in faith.—*Life's Evening.*

AND thus, O slothful heart of mine! if thou wert also found
Dauntless in labor for thy Lord, though dreariness abound,
Linked to his heart with bands of love, by life or death unriven,
Thou, too, wouldst wait for dying grace, and "live in sight of heaven."

THE LOVING DISCIPLINE OF PAIN.

GOD now inquires whether you are truly his child — whether, in full view of the rod that is raised, you will say, "It is the Lord, let him do what seemeth him good"? God is now applying a test, that you may know whether you are truly such He has placed you in the alembic of suffering. It may seem to you that in the process there is intensity, and even fury. But all that he does is needful. It is not in ANGER that the refiner puts the precious metal into the fire. God knows infinitely well what is best for you. Your physician may mistake your case; but God never. Nothing comes from him that betrays want of skill, or that proves pernicious. Take then this suffering as a paternal dispensation, and bless God that he has ordered it.

WHAT, many times I musing asked, is man,
If grief and care
Keep far from him? he knows not what he can,
What can not bear.

He, till the fire hath purged him, doth remain
Mixed all with dross:
To lack the loving discipline of pain
Were endless loss.

Nay, then, but He who best doth understand
Both what we need
And what can bear, did take my case in hand,
Not crying heed.

THE EARLY DEAD.

WE weep for the dead. Let nature speak, and we should all say that we do well to weep for them, especially when death comes suddenly upon them in the days of their youth. Oh! what a strange and melancholy change have they experienced! Instead of the cheerful light of day, the unbroken darkness of the grave covers them forever! They are alone, solitary there; their only companion is the worm. All their earthly hopes have died — all their expectations have perished.—*Charles Bradley.*

ONE year ago—what loves, what schemes
   Far into life!
What joyous hopes, what high resolves,
   What arduous strife!

No note, no hush of merry birds
   That sing above,
Tell us how coldly sleeps below
   The form we love!

THE NARROW STREAM OF DEATH.

IN a few hours after she was attacked it became evident to those around her, and to herself, that the mortal blow had been struck. She needed no one to tell her of it; she felt within herself that life was fast ebbing away, and said of the weariness upon her, that it must

be the weariness of death. When a friend who stood
by her expressed her sorrow that she should take such
a view of her case, she said : "I submit to His will, and
desire that he may do with me as seemeth to him good ;
though it is very painful to be separated from my dear
husband and my sweet children. But I commit them
all into the hands of my Saviour. It will be a short
separation, and then we shall meet to part no more."
Being asked if she felt afraid to die, she replied : "No;
I had always expected that the prospect of death would
almost frighten me out of existence ; but now it has no
terrors. I rely on Jesus, and feel I shall be happy when
I die. It is better to depart and be with him, where I
shall be completely freed from sin."

Once, with a sweet expression of countenance, she
said : "How much is implied in those words : 'The peace
of God, which passeth all understanding!'" Much on
her lips, and more in her thoughts, was that name —
name above every name—Jesus. Among her prayers
to him were : " O Lord Jesus! place underneath me thy
everlasting arms! Jesus, receive my spirit. O Lord
Jesus! receive me on the other side of Jordan." Nor
did her heart spend its emotions in prayer alone ; it
was attuned to praise. She said : "I want a hymn
sung." "What hymn?" "The hymn about crossing
over Jordan," she said ; and it was sung ; and soon after
she crossed the stream — the narrow stream of death.
Nor did Jesus wait for her on Canaan's bright side of
the stream ; but he came over to earth's dark shore of
it, and himself took her across That stream must be
narrow, it was so soon passed ; and all was so calm,

there could not have been a ripple on its surface. O
death! where was thy sting? O grave! a feeble, fearful female, with only a few hours to arm herself for the
conflict, and to take leave of her babes, met thee, and
was more than victor through Him who gave her the
victory.— *William Nevins.*

> ALONE? ah! no—in closer grasp than mother's fondest hold,
> The Lord of life and death received that soul to bliss untold.
> There was no need of human help when Christ could ease the chill,
> And gently touch the throbbing heart, and bid the pulse be still.
>
> Bright is the sunset splendor thrown from many a dying-bed,
> And eloquent the influence of all the saintly dead:
> Far down the turbid waves of time, those rays will burn and beam,
> As lighted pinnace launched by night on Oriental stream.

THE CHILD-ANGEL.

RISING up after her long vigil, she went noiselessly
down-stairs toward the room where her child slept the
last long sleep. As she was entering, a voice struck
her ear, as if some long-remembered music had just
sounded; the chord vibrated against her heart. She
paused; the voice asked for Antoinette—little Antoinette Hayden—and another voice mournfully murmured
the sad truth. " Dead!" exclaimed the stranger —
" little angel dead!"

And then came feet along the passage, and a tall man
stood before Mrs. Hayden. " You do not know me,
Mrs. Hayden," he said, as after a moment striving to
possess his self-command, he spoke.

"I do not, indeed," replied the bereaved mother, in low tones.

"Ah! my dear madam, I am he whom your child's artless questions, morning after morning, pierced to the heart; I am poor Loose Ben. Day and night have the lovely features of that angel-child been before my vision. Every morning the sweet, clear tones have sounded on my ear, 'Does you love God?' and, oh! I have come to find her in heaven." He bowed his head and wept, then softly followed the mourning mother into the shaded parlor. Death had not stolen one line of beauty from that heavenly face — it was lovely in spite of death.

"O Antoinette! dear little Antoinette!" sobbed the strong man; "you found me in ignorance, and blessed me with those holy hands. They were the first pure fingers that touched me with the touch of love, and made my buried heart throb with new life. O little Antoinette! you were the first one to lead me to my Saviour; on your infant breath my name was first carried to Christ. O my lamb! canst thou not look down upon me, and see me bend over thee, blessing even thy inanimate clay? But the tomb can not hold thee, infant disciple. Already is she up there! The brightness of the glory, O Lord God of hosts! falls upon her temples. She hath led souls to thee, mighty Redeemer, and thou wilt give her a crown of life."

He ceased and bowed his head upon the coffin. He had been converted through her ministrations, and since his entrance into the Gospel ministry, he counted those who believed in Jesus, through his faith and his ministry, by hundreds; and he laid his trophies, in the name of Jesus, before the gentle child who had taught him Christ.

Reader, I have not written fiction. The dust of the child has slept in the green graveyard where the flowers are springing to-day, twenty-three years. Twenty-three years she has been a seraph in glory. Twenty-three years she has looked upon Jesus, her Saviour and her Redeemer. Oh! what do you and I seem beside this beautiful seraph? Though we drink of the fountain of earthly wisdom, we can not attain to a tithe of that divine knowledge that fills her cup of bliss this day. Twenty-three years in the presence of the Lord of life, going up and down the steps of light — walking and talking with angels — pure, consecrated, holy.

'Tis ever thus, 'tis ever thus, with all that's best below,
The dearest, noblest, loveliest are always first to go!
The bird that sings the sweetest, the vine that crowns the rock,
The glory of the garden, the flower of the flock.

'Tis ever thus, 'tis ever thus, with creatures heavenly fair,
Too finely formed to bide the brunt more earthly natures bear;
A little while they dwell with us, blest ministers of love,
Then spread the wings we had not seen, and seek their home above!

OUR IGNORANCE OF THE FUTURE.

OUR ignorance of the future brings our best-laid schemes to ruin; our ruined schemes tell us of our dependence on the world's great Master; we are reminded of a forgotten God. And here is your consolation : "The Lord knoweth the way that you take." "He knoweth thy walking through this great wilderness." He foresees all that is coming on you in it, and he has provided for all ; yes, he provided for every want and sorrow you can ever know, before you came into being, and has left you nothing to care about but this, " to win Christ and be found in him ;" to lay hold of his salvation ; to hold fast by him for a few short, stormy years, and then to enter into everlasting joy. Look forward you may, but let it not be into the low, dark valley of uncertainties that lies immediately before you —a confused, misty scene you can not penetrate ; look over it. Lift up your eyes to the bright hills that rise beyond it. There they are, resting on their everlasting foundations ; and, oh! the blessedness of even a distant glimpse of them! We no longer heed then the valley's darkness, or the valley's roughness. We rather say, " There is light, there is rest, there is heaven before us ;" and go on our way rejoicing.— *Charles Bradley.*

WHAT shall the future progress be
    Of life with me ?
God knows—I roll on him my care—
Night is not night if he be there.
When daylight is no longer mine,
And stars forbidden are to shine,
    I'll turn my eyes
To where eternal day shall rise.

That coming light no gloomy cloud
    Can quite enshroud !

Through all our doubts—above the range
Of every fear, and every change—
My faith can see, with weary eye
The dawn of heaven on earth's dim sky,
    And from afar
Shines on my soul the morning star.

REST.

"And he said unto Jesus, Lord, remember me when thou comest into thy kingdom. And Jesus said unto him, Verily I say unto thee, to-day shalt thou be with me in Paradise."

That prayer and its acceptance make our hearts thrill, even in our coldest moments, with longings for the same assurance. Rest and safety! and with Him! But fully to appreciate the rest of Paradise, we must understand and realize the unrest of earth; and this, perhaps, is what few do. There is rest to the heavy-laden with sin, in the sense of a Saviour's forgiveness; there is a calm to the wearied spirit, when it looks up in loving confidence to an Almighty Protector; but with all—in, about, and inseparably connected with all —is the sleepless and abiding sense of danger.

In Paradise is no danger; therefore in Paradise alone is there rest. "To-day shalt thou be with me." Can it be possible? To-day, with its cares, its business, its its projects, thoughts for others, fears for them, fears also for ourselves! To-day! with its anxious, wandering prayers, its hasty meditations, its weak struggles, its humiliating defeats, its far-reaching anticipations of

greater failures; this very day, may there indeed be
rest? Lord, teach us to long for it. Teach us to
yearn for that unspeakable calm, that perfect, untrou-
bled safety!—*Sewell.*

On, blest immortal, on, through boundless space,
And stand with thy Redeemer, face to face;
 And stand before thy God!
 Life's weary work is o'er;
 Thou art of earth no more;
No more art trammeled by the oppressive clay:
 Thou art a welcome guest;
 This city's name is Rest;
 There shall no fear appall,
 Here love is all in all;
Here shalt thou win thy ardent soul's desire;
Here clothe thee in thy beautiful attire.
 Lift, lift thy wondering eyes!
 Yonder is Paradise,
 And this fair, shining band
 Are spirits from that land!
And those who throng to meet thee are thy kin,
Who have awaited thee redeemed from sin!
The city's gates unfold—enter and rest within.

WAITING IN HOPE.

I MOST willingly forsake this world, this vexatious,
troublesome world, in which I have no other business
but to rid my soul from sin; with patience and courage

bear my eminent misfortunes, and ever hereafter be above the smiles and frowns of it. Those are happy who in the midst of confusions can faithfully believe the end of all shall be rest ; spiritual joy will grapple with earthly griefs, and so far overcome as to give some tranquillity to a mind tossed to and fro, as mine has been, with the evils of this life. I am much encouraged by your allowing that I have a just sense of sorrow ; it excites me better to struggle for my duty, doing all I can ; and I hope my duty shall always prevail above the strongest inclination. I believe to assist my yet helpless children, is my business ; which makes me do many things, the performance of which is hard enough to a heavy and weary mind ; and yet I bless God I do it. Indeed, Doctor, you are extremely in the right to think that my life has been so imbittered ; it is now a very poor thing to me ; yet I find myself careful enough for it. I think I am useful to my children, and would endure hard things, to do for them till they can do for themselves. The pensive quiet I hope for here, I think will be very grateful to my wearied body and mind ; yet when I contemplate the fruits of the trial and labor of these last six months, it brings some comfort to my mind, as an evidence that I do not live only to lament my misfortunes, and be humbled by those heavy chastisements I have felt, and must forever in this life press me sorely. My glass runs low : the world does not want me, nor I want that ; my business is at home, and within a narrow compass. We must wait our day of consolation till this world passes away ; an unkind and trustless world it has been to us. Why it has been such, God knows best; all his dispensations are beautiful and must be good, and good to every one of us, and even these dismal ones, if we can bear evidence to our own souls that we are better for our afflictions; though my eyes are ever ready to pour out marks of a sorrowful heart, which I shall carry to the grave, that quiet bed of rest.—*Lady Rachel Russell.*

> Two hands to work addrest,
> Aye for his praise;
> Two feet that never rest,
> Walking his ways;
> Two lips still breathing love,
> Not wrath nor fears;
> Two eyes that look above,
> Through all their tears!

"STRONG IN CHRIST."

"I AM not tired of my work, neither am I tired of the world; yet when Christ calls me home, I shall go with the gladness of a boy bounding away from his school. Perhaps I feel something like the young bride, when she contemplates resigning the pleasant associations of her childhood for a yet dearer home — though only a very little like her, for THERE IS NO DOUBT RESTING ON MY FUTURE." "Then death would not take you by surprise," I remarked, "if it should come even before you could get on board ship?" "Oh! no," he said; "death will never take me by surprise — do not be

afraid of that—I feel so STRONG IN CHRIST. He has
not led me so tenderly thus far, to forsake me at the
very gate of heaven."—*Emily C. Judson.*

Our ransomed dead, who clasped the Cross in dying
With else despairing clutch;
And felt a strong Right Arm beneath them lying,
His, whom they loved so much!

BEREAVEMENT.

I LEFT papa soon, and went into the dining-room. I
shut the door; I tried to be glad that I was come home.
I have always been glad before, except once; even then
I was cheered. But this time joy was not to be the
sensation. I felt that the house was all silent — the
rooms were all empty. I remembered where the three
were laid — in what narrow, dark dwellings — never
more to reäppear on earth. So the sense of desolation
and bitterness took possession of me. The agony that
WAS TO BE UNDERGONE, and WAS NOT to be avoided,
came on. I underwent it, and passed a dreary evening
and night, and a mournful morrow. Sometimes when
I wake in the morning, and know that solitude, remembrance, and longing are to be almost my sole companions all day through; that at night I shall go to bed
with them; that they will long keep me sleepless; that
next morning I shall wake to them again, sometimes I
have a heavy heart of it. But crushed I am not yet,
nor robbed of elasticity, nor of hope, nor quite of en-

deavor. I have some strength left to fight the battle of life. I am aware, and can acknowledge, I have many comforts, many mercies. Still I can GET ON. But I do hope and pray, that never may you, or any one I love, be placed as I am. To sit in a lonely room —the clock ticking loud through a still house, and have before the mind's eye the record of the last year, with its shocks, sufferings, losses — is a trial. — *Charlotte Brontë.*

SLIGHT are the causes, frail, unfeared,
    That desolation bring;
Shrines through a lifetime's toil upreared
    One day may downward fling;
And still the shell of home be there—
    The void within, how bleak and drear!

'Tis through His will the homes we love
Are rifled. There is a safer, holier fane!
Its glory no assault may stain.
Why stand we gazing here on vacant niche,
When angels show the home, beyond imagining rich?

DEATH OF A HUSBAND.

I CAN not tell you, dear mother, in what state I am since the fatal month has commenced. It is two years to-day since we departed for Plombières. During all the journey he loaded me with attention and testimonials of his affection. Each hour, alas! has its sweet remembrance, and each hour brings me nearer the terrible day on which I lost so much. How falsely men judge when they think time will heal wounds! Grief is no longer so devouring, but it is not less intense; the more the wound seems to heal upon the surface, the deeper also becomes the suffering. I suffered a thousand deaths, and was fearfully depressed, till at the grave I again found the Lord. Now I am at peace with him, with my cross, with my future upon earth. Thank God for me; he has wonderfully sustained me; he has granted me his peace, his presence; he has strengthened and revived my poor, withered, stricken heart.

I have been obliged to receive the ministers and royal household at Paris; the reception was in the evening, in the very apartments where HE appeared so often. They were brilliantly lighted as on former occasions, and presented the aspect of a fête; but alas! what a fête. In the midst of the crowd there was but one thought, one regret; above all the surrounding group there arose the noble, cherished portrait of the Prince.—*Helen, Duchess of Orleans.*

THE silent picture on the wall,
　The burial-stone
Of all that beauty, life, and joy,
　Remain alone!

One year, one year, one little year,
And *so much gone!*
And yet the even-tide of life
　Moves calmly on.

The grass grows green, the flowers bloom fair,
  Above that head ;
No sorrowing tint of leaf or spray
  Tells he is dead !

Lord of the living and the dead,
  Our Saviour dear,
We lay in silence at thy feet
  This sad, sad year.

THE OLD HOME.

IN that moment of collapse the spirit of little George had escaped from the form that held it, leaving it to all appearance uninjured ! The soul had leaped upward to the bosom of the angel of the covenant, and long before the other bodies, then apparently lifeless as his, had agonized back into life, his peaceful remains were laid in a soft wrapping-rug on the green grass-bank, and he had taken in the first draught of immortality.

.   .   .   .

Permission being given by the physicians for us to have one look at Freddy, he was carried down-stairs on a small mattrass. Room was made across our feet, and he lay there so sweet and bright-looking, with his eyes half-raised, so little changed from that last look on the railway-bank, lovelier than he had almost ever looked before, that we could not believe he was uncon-scious. It was only when the physicians had raised us in bed to kiss him, and taking his hand, we asked him if he did not know us, that we saw that he was already deaf to all earthly voices, and that his time was counted by seconds. His papa prayed for him and gave him up to God. "O Lord! thou hast heard his earnest cries for a new heart, and to be washed in the blood of Jesus, and taken to heaven when he died. Answer them all, and take him to thyself." He was carried away, and expired in a few moments. A Sabbath sun had lighted him home; and oh! how much of our poor hearts went with him!

Frederic and George were laid to rest in their infant brother's grave. We have lent them to the Lord, and it depends on us whether we are totally separated from them or not. It is our fault if the wilderness-path does not often border on the spirit-land. "IF YE LOVE ME, YE WOULD REJOICE." Like a soft, solemn chime of far-off bells, these words rung through our empty hearts the last hours of our railway journey back home. You can hardly imagine what a changed house was ours on our return. Sweet still, for their sakes, is all they have left behind them. Fragrant are the flowers they planted, and the garden-trees that shadowed them. Perfumed the rooms they lived and prayed in — chosen spots now every one of them. The silence of them may seem terrible, but praise can break it; and where should survivors be able to get so clear a view of the new home whither the absent ones are gone, as from the place that once knew them so well? There are pleasant memories clinging to its walls that can not grow in any other scene.—*The Way Home.*

THE old house by the lindens
Stood silent in the shade;
And on the graveled pathway
The light and shadow played.

I saw the nursery-windows
Wide open to the air—
But the faces of the children
They were no longer there.

The large Newfoundland house-dog
Was standing by the door;
He looked for his little playmates
Who would return no more.

They walked not under the lindens,
They played not in the hall;
But shadow and silence and sadness
Were hanging over all.

The birds sang in the branches
With sweet, familiar tone;
But the voices of the children
Will be heard in dreams alone.

WHERE ARE OURS NOW?

So the light in your dwelling has gone out, my poor brother, and it is all darkness there, only as you draw down by faith some faint gleams of the light of heaven; and coldness has gathered round your hearthstone; your house is desolate, your children are scattered, and **you a homeless wanderer** over the face of the land.

We have both tasted of these bitter cups once and
again ; we have found them bitter, and we have found
them sweet, too. Every cup stirred by the finger of
God becomes sweet to the humble believer. Do you
remember how our late wives and sister Stevens used
to cluster round the well-curb in the mission inclosure
at the close of day ? I can almost see them sitting
there, with their smiling faces, as I look out of the
window at which I am now writing. Where are ours
now ? Clustering around the well-curb of the fountain
of living waters, to which the Lamb of heaven shows
them the way ; reposing in the arms of Infinite Love,
who wipes away all their tears with his own hand. Let
us travel on and look up. We shall soon be there.
As sure as I write or you read these lines, we shall
soon be there. Many a weary step we may yet have
to take ; but we shall get there at last : and the longer
and more tedious the way, the sweeter will be our repose.—*Adoniram Judson.*

> FAIN, till His love the flow of anguish stanches,
>   When our beloved flee,
> Fain would we follow where each frail raft launches
>   Far on the eternal sea.
>
> Fain would we hear their new-found joy outgushing
>   In heaven's triumphing psalms.
> And feel a fragrance round our foreheads rushing,
>   Fanned from their deathless palms.
>
> O friend ! our Father doubtless hath fair gardens,
>   Beyond the walls we see ;
> With restful glades and souls we love for wardens,
>   But He still keeps the key.

OVERRULING PROVIDENCE.

"If thou hadst been here, my brother had not died."
These little words plainly show that these afflicted sisters both believed that, had they been permitted to
order the course of events, the result would have been
far happier. If something had happened which has
not happened, the event might have been less wretched.
Oh! how often do reflections similar to this barb the
arrow of affliction with a poignancy which nothing
else can give! These are the thoughts which in our
wretchedness make us doubly wretched: "If we had
taken such a course, if we had acted in some other
manner, how different would have been the issue!"
There can be nothing more unwise, perhaps few things
more unholy, than reasoning thus. In dwelling upon
secondary causes, we overlook the first great cause of
all—the God of heaven and earth, who alone ordereth
all things, and doeth all things well. Has the Lord no
share in the decision? Did he not direct our present
disappointment? Was he not present when our friend
was taken from us? Duties are ours, events are God's.
—*Blunt.*

> One adequate support
> For the calamities of mortal life
> Exists—one only—an assured belief,
> That the procession of our fate, howe'er
> Sad or disturbed, is ordered by a Being
> Of infinite benevolence and power,
> Whose everlasting purposes embrace
> All accidents, converting them to *good.*

OUR EARLY LOST.

YES, blessed Saviour, in thy bosom nestles the lamb of our fold. We can not think of him without remembering thy sweet words: "Suffer the little children to come unto me."

It is not, then, the illusion of fancy, it is the dictate of Christian faith, to look toward the holy city, and, within its gates of pearl, to see the little one that has been taken from us, now a pure, beautiful spirit, robed in celestial beauty, with a crown on his head, and a harp in his hand, beckoning us to come up hither.

Oh! it was sweet to hear his voice in the glee of infancy; sweet to feel his lips pressed to ours; sweet to listen to his infant prayer, or gentle murmur, when we hummed the evening lullaby. But he is brighter, fairer, happier there; and we shall soon rejoin him in our Father's house, a reünited family, all the more blessed because we have been for a little while separated, and then we shall part no more forever. It is a blessed thought, that when one of our children dies in infancy, it sleeps in Jesus. We are *sure* of one in heaven. The rest may grow up in sin, and die in sin, and be lost, but one is safe. They only can be said to possess a child forever who have lost one in infancy.—*S. I. Prime.*

> My lambs! I loved them so
> That when the elder Shepherd of the fold
> Came, covered with the storm, and pale and cold,
> And begged for one of my sweet lambs to hold,
> I bade him go.

He claimed the pet;
A little fondling thing, that to my breast
Clung always, either in quiet or unrest;
I thought of all my lambs I loved him best;
    And yet—and yet,

    I laid him down
In those white, shrouded arms, with bitter tears,
For some voice told me that, in after-years,
He should know naught of passion, grief, or fears,
    As *I* had known.

    And.yet again
The elder Shepherd came; my heart grew faint;
He claimed another lamb, with sadder plaint—
Another! she who gentle as a saint,
    Ne'er gave me pain.

    " Is it thy will?
My Father, say, must this pet lamb be given?
Oh! thou hast many such, dear Lord, in heaven;"
And a soft voice said : " Nobly hast thou striven,
    But peace—be still."

    Oh! how I wept,
And clasped her to my bosom with a wild
And yearning love! my lamb, my pleasant child —
Her, too, I gave; the little angel smiled
    And slept.

  .   .   .   .   .   .

I sit and think, and wonder, too, sometime
How it will seem, when in that happier clime,
It never will ring out the funeral-chime
    Over the dead!

DEATH OF A SON.

IT was not, therefore, without some small degree of surprise that, at eight o'clock in the evening, we perceived it evident that he was sinking very fast. His three or four immediate relatives, the physician, and the old affectionate servants were assembled in the room, and he spoke continuously for a considerable time, with apparently little difficulty of utterance, and with the most perfect composure and command of mind and language; addressing or adverting to each of us, expressing a grateful sense of the kindness he had experienced; his request to be forgiven any thing in which he had ever been blamable toward any of us; his wish that each one might receive one more religious admonition from his death; his trust that we shall all meet again in a happier world; and his hope in the divine mercy through Jesus Christ. He was sensible till within the last hour. When I thought his mind was finally withdrawn from us, and his eyes finally closed, I touched his face, and spoke to him, and he instantly looked up, and, with evident intelligence, spoke one word in reply; and a few moments after, looking at his mother, he in an affectionate tone said, " Mamma!" the last word he uttered. A little after, he sunk in sleep, and passed from sleep into death. In looking on the deserted countenance, through which mind and thought had so recently, but as it were a few minutes before emanated, I felt what profound mystery there was in the change. What is it that has gone? what is it now? Thus there is a termination of all the cares, solicitudes, and apprehensive anticipations concerning our son and your pupil. He is saved from entering on a scene of infinite corruptions, temptations, and grievances, and borne, I trust, to that happy region where he can no more sin, suffer or die; safe and pure and happy forever! In such a view and confidence I am (and my wife, too, though for the present more painfully affected) *more* than resigned to the dispensation; the consolation greatly exceeds the grief. Indeed, I believe that to me the consolatory considerations have much less to combat with than in the case of parents generally. Probably I may have expressed to you, that I have such a horror of this world, as a scene for young persons to be cast and hazarded into, that habitually, and with a strong and pointed sentiment, I congratulate children and young persons on being intercepted by death at the entrance into it, except in a few particular instances of extraordinary promise for piety, talent, and usefulness. If, as in *our* case, parents see their children, in an early period of life, visited by a dispensation which, in *one and the same act, raises them to piety and dooms them to die, so that they receive an immortal blessing at the price of death,* oh! methinks it is a cheap cost, both to them and to those who lose them! In one of my first conversations with John on his irrecoverable situation, when I said, "We shall be very sorry to lose you, John," he calmly and affectionately replied: "You will not be sorry, if you have cause to believe that I am beyond all sorrow."—*John Foster.*

. . . . . .
Youth's brightest hopes decay,
Pass like morn's gems away,
Too fair on earth to stay
Where all is fleeting.

When in their lonely bed,
Loved ones are lying;
When joyful wings we spread
To heaven flying,
Would we to sin and pain
Call back their souls again,
Weave round their hearts the chain
Severed in dying?

DEATH OF A YOUNG SOLDIER.

WITH Captain Hammond's name you will be fami
liar. A braver soldier never on that day mounted the
Redan. A Christian of more unaffected piety never
entered the presence of God. He had only been in
the Crimea forty-eight hours when he was killed.
When the Rifles were forming for the attack, a
young subaltern, going into action for the first time,
who had come out with Captain Hammond, addressed
him : "Captain Hammond, how fortunate we are, we
are just in time for Sebastopol." Hammond's eyes
were gazing where the rays of the sun made a path of
golden light over the sea, and his answer was short and
remarkable, and accompanied by the quiet smile which
those who knew him will so well remember. "*I am
quite ready,*" said he. The next that was seen of him
was, when his sword was flashing above one of the em-

brasures of the Redan. Pressing forward then himself into the heart of the work, with a color-sergeant
and one or two devoted men who had bound up their
fate in his, his sword is seen flashing far in advance in
personal encounter. Once or twice in that deadly fray,
his form appears through the embrasures; and for a
few moments before his strong arm the Russian foe-
man retires and closes again. But to *him* neither
earthly crown, nor medal, nor grateful country's praise
is in store for these moments of devotion. The deadly
bayonets close around him, the sword drops from the
uplifted hand, and he sinks into the arms of an officer
of the Forty-first. But with angels and seraphs and
the host of heaven, who were waiting "on the other
side of the river," there were hymns of joy that day.
"Eye hath not seen, nor ear heard, neither hath it entered the heart of man, the things that God hath prepared" for that happy, ransomed spirit. Before night
an effort was made to recover the body. Capt. R——,
an officer of the Seventy-second Highlanders, at much
risk, took with him a party of men, and made search
in vain. In the morning, very early, a party of rifle-
men approached from the works toward the camp.
The precious object of their search had been found.
An expression of sweet peace rested on the placid
features. A very small puncture, close to the heart,
told how instantaneous must have been his death. Almost upon the wound, a locket, bathed in his heart's
blood, was lying.

The following extract from Captain Hammond's last
letter to his wife, written on the morning of the day
of his death, will be read with mournful interest:

"The order for the attack has just come out; thankful I am that you can not know it, dearest, beforehand. F—— with a hundred men form the covering party to the whole. The remainder of our battalion form part of the reserve, and follow up the attack. The Lord Jesus be with you!

"P. S.—6.30 A.M. I have had a peaceful time for prayer, and have committed the keeping of my soul and body to the Lord my God, and have commended to his grace and care my wife and child, my parents, brothers, and sisters, and all dear to me. Come what will, all is well. This day will be a. memorable one. Farewell, once more! Psalm 91 : 15 is my text for to-day, especially the words: 'I will be with him in trouble!' "—*Life of Captain Hammond.*

> Go to the grave in all thy glorious prime,
>   In full activity of zeal and power ;
> A Christian can not die before his time :
>   The Lord's appointment is the servant's hour.
>
> Go to the grave, at noon from labor cease ;
>   Rest on thy sheaves, the harvest task is done ;
> Come from the heat of battle, and in peace,
>   Soldier, go home ; with thee the fight is won.
>
> Go to the grave, for there the Saviour lay
>   In death's embrace, ere he arose on high ;
> And all the ransomed, by that narrow way,
>   Pass to eternal life beyond the sky.
>
> Go to the grave—no, take thy seat above ;
>   Be thy pure spirit present with the Lord ;
> Where thou for faith and hope hast perfect love,
>   And open vision for the written word.

INDEX.

www.ingramcontent.com/pod-product-compliance
Lightning Source LLC
Chambersburg PA
CBHW021812190326
41518CB00007B/566